Cosmos Proof of the Existence of God

A Physical Proof of God
Cosmos Quantum Fractal Curve
Experiment Proves Cosmos Theory
Divinity and the Cosmos
God as Initiator of the Cosmos
Physics of Predestination & Free Will
God as the Ultimate Quantum Observer
The Primordial Cosmos Point

Stephen Blaha Ph. D.
Blaha Research

MMXXVI

.

ISBN: 979-8-9986071-8-9

Rev. 00/00/01 February 10, 2026

To Margaret

Some Other Books by Stephen Blaha

SuperCivilizations: Civilizations as Superorganisms (McMann-Fisher Publishing, Auburn, NH, 2010)

Unification of God Theory and Unified SuperStandard Model THIRD EDITION (Pingree Hill Publishing, Auburn, NH, 2018).

The Exact QED Calculation of the Fine Structure Constant Implies ALL 4D Universes have the Same Physics/Life Prospects (Pingree Hill Publishing, Auburn, NH, 2019).

Passing Through Nature to Eternity ProtoCosmos, HyperCosmos, Unified SuperStandard Theory (Pingree Hill Publishing, Auburn, NH, 2022).

HyperCosmos Fractionation and Fundamental Reference Frame Based Unification: Particle Inner Space Basis of Parton and Dual Resonance Models (Pingree Hill Publishing, Auburn, NH, 2022).

The Cosmic Panorama: ProtoCosmos, HyperCosmos, Unified SuperStandard Theory (UST) Derivation (Pingree Hill Publishing, Auburn, NH, 2022).

God and and Cosmos Theory (Pingree Hill Publishing, Auburn, NH, 2023).

Newton's Apple is Now The Fermion (Pingree Hill Publishing, Auburn, NH, 2023).

Cosmos Theory: The Sub-Particle Gambol Model (Pingree Hill Publishing, Auburn, NH, 2023).

Cosmos-Universe-Particle-Gambol Theory (Pingree Hill Publishing, Auburn, NH, 2024).

Structure and Dynamics of Cosmos Theory and the Unified SuperStandard Theory (Pingree Hill Publishing, Auburn, NH, 2024).

Black Holes, White Holes, and Superluminal Starship (Pingree Hill Publishing, Auburn, NH, 2025).

The Sequence Theory Foundation of Cosmos Theory and Gambol Theory (Pingree Hill Publishing, Auburn, NH, 2025).

BEYOND TIME AND SPACE Fermion Strata & Consistency Conditions for Sequences, Cosmos Spaces and Gambols (Pingree Hill Publishing, Auburn, NH, 2025).

The C o s m o s as a Physics Organism (Blaha Research, Auburn, NH, 2025)

The Root of Reality (Blaha Research, Auburn, NH, 2025)

Fractal Cosmos (Blaha Research, Auburn, NH, 2025)

Fractal Cosmos II (Blaha Research, Auburn, NH, 2025)

Fractal Cosmos III (Blaha Research, Auburn, NH, 2025)

Fractal Cosmos IV (Blaha Research, Auburn, NH, 2025)

The Entropic Basis of Cosmos Theory The Agent of Origin (Blaha Research, Auburn, NH, 2025)

Available on bn.com, Amazon.com, Amazon.co.uk and other international web sites as well as at better bookstores.

CONTENTS

FIGURES and TABLES

Introduction

Thomas Aquinas gave two proofs of the existence of God based on Aristotelian Logic. This book presents a proof based on the Physical considerations of Cosmos Theory. It begins with an outline of Cosmos Theory structure and its substantial verification based on known Physics Experiment results in our universe. The book shows Cosmos Theory has a strong experimental basis in known Particle Physics. This basis follows directly from the features of Cosmos Theory. Therefore, despite the view of some that one cannot prove that Cosmos Theory describes known Physics, the book shows a direct connection to known Physical results including the Standard Model of Particles.

For our universe, the known results include: the numeric values of all known coupling constants, the set of internal symmetries – the Strong interactions and the ElectroWeak interactions as well as General Relativity, the spectrum of the known fundamental fermions and quark and lepton masses, and, most importantly, the structuring of the internal symmetries and fermion masses into sequences. In addition the parameter n of a Yang-Mills SU(n) is shown to fully determine the corresponding coupling constant for a universe of any dimensions. Similarly the ElectroWeak Weinberg angle is also determined by Cosmos Theory. These features are enhanced by the additional proposed interactions and fermions of the Unified SuperStandard Theory including fermion strata and fermion gambols.

Cosmos Theory has spaces that describe other possible universes of 4 or more dimensions. These universes have analogous features. All Physics calculations in these universes are finite – no infinities – due to the author's Two Tier form of Quantum Theory. This book considers features of Parent and child universes such as mass and energy.

The book describes the primordial point from which Cosmos Theory springs. Is there anything before the primordial point? The book shows that any proposed prior "space" or "universe" only leads to an unnecessary eternal recurrence of past "Cosmoses."

From where does the primordial point arise? The only viable solution is the existence of an Ultimate Quantum Observer and creator: God. This solution eliminates the paradox of the Schrödinger Cat. If no visible observer is present, then God determines the aliveness of the cat. No paradox. So much also for "One Hand Clapping" paradoxes.

The Cosmos evolves from the primordial point in a manner the author described in previous work and presents here.

The book describes aspects of God's Physical roles. In particular it shows that the primeval Cosmos component, the Cosmos Fractal Curve, can be Quantum only because the ultimate Observer is present. The Cosmos Quantum Fractal Curve is derived in detail.

This book desiccates the Cosmos components one by one. Realizing that no time exists, it proceeds to couple them instantaneously under the umbrella of the Ultimate Observer.

When? Endlessly. It supports the view that the Deity is present in all things and all spaces and times – based on a Physics perspective, The aspects of God's interaction with Reality are described using the author's suggestion that Reality can be "tweaked" in such a manner as to conform with Physics Laws, which, after all, He formulated – and yet still accomplish God's work.

1. Where Shall We Begin?

In the "Beginning" all structural Cosmos components (features) were simultaneously present given the absence of Time: the Primordial Point, the Cosmos Fractal Curve, the Extended Master Equation, the Consistency Condition, the spectrum of Cosmos spaces, the Parent universe, and the Parent universe HyperSphere as exemplified in in Fig. 1.1.

To describe Cosmos Theory we must separate the components of Cosmos Theory sequentially into a series of stages. The goal is to describe these features so as to show their interrelations. The structure of the Cosmos Theory features coexist simultaneous and eternally without beginning or end. Having determined the structure of the Cosmos we can consider the nature of universes. One universe, the Parent universe (of all child universes), stands out since it coexists with the structure of the Cosmos. The Parent universe exists simultaneously with the Cosmos structure components. It has no beginning and no end although we use a Consistency Condition to establish its existence and its nature upon its formation (knowing that the Consistency Condition is a necessary condition that applies forever for the Parent universe.)

The child universes are subject to time since they exist within the Parent universe and thus have a beginning or end? in Parent universe time. Within each child universe a time exists. This time may be related to Parent universe time in principle although it cannot be determined experimentally. Each child universe begins with a certain energy that is well defined due to its presence within the Parent universe. (Some General Relativistic quibbles over the uncertainty of determining energy are removed by the child universes presence within a larger universe.)

We begin with an overview of the stages of Cosmos Theory structuring. Later chapters will consider these stages in detail.

1.1 The Beginning

We assume the initial state of the Cosmos was a single mathematical point with no structure or properties. From this point a space-filling fractal curve is generated. (Chapter 7.) The resulting unit square is then made continuous and scaled upwards to a complete two dimension space. We introduce a time dimension as the distance along the fractal curve in the steps followed in its construction. The completed three dimension space is called the *Cosmos Space*.

1.2 Master Equation and Cosmos Spaces

We can define a Quantum dynamic equation in the Cosmos Space. This equation has an energy which can be put in the form:

$$E_n = 2^n$$

which corresponds to spaces in Cosmos Theory. The power n is related to the space-time dimension r of the space. For one set of spaces called HyperCosmos[1] spaces n = r + 4. For the other set of spaces, Second Kind[2] HyperCosmos spaces n = r + 3. The space-time dimension is restricted to even values of r to avoid double-counting.[3]

The "energy" E_n is the number of dimensions in the dimension array of the space. (We associate an array of dimensions with each space.)

Having defined Cosmos spaces and their dimension arrays we next proceed to define universes. The process of defining universes begins with determining the Parent universe, which is populated with child universes. See below.

1.3 Cosmos Spaces

HyperCosmos spaces can have their dimension arrays placed in the form of a square array with half of the dimensions for Normal fermions and symmetries, and half for Dark fermions and symmetries.

Second Kind HyperCosmos spaces can have their dimension arrays placed in the form of a rectangular array with all dimensions *only* for Normal fermions and symmetries.

The difference between these two types of spaces, which appear with equal space-time dimensions, enables us to define Quantum states that are superpositions of the two types:

$$|\text{space of dimension } r\rangle = \cos(\theta)|\text{HyperCosmos space}\rangle + \sin(\theta)|2^{nd} \text{ Kind HyperCosmos space}\rangle$$

where θ is an angle. Based on this superposition we can map the spaces to a double spiral helix reminiscent of DNA as shown in Fig. 1.1 using variations in θ.

The sets of HyperCosmos spaces are extended to generate the unification spaces shown in Fig. 1.6. These spaces are used to embody "rotations" among the sets of symmetries corresponding to the components of dimension arrays. Thus the entire set of symmetries of a space (and universe) can be transformed amongst themselves. As a result one obtains a fully unified set of symmetries (including symmetry breaking interactions.)

Unification uses a type of reference frame called a Fundamental Reference Frame (FRF). One can transform the dimensions of a dimension array (and its maps to fermions and symmetries) to one non-zero dimension in an FRF. The transformations are combinations of General Relativity – Internal Symmetry transformations. The inverses of these transformations from one non-zero dimension to a filled dimension array are also possible. FRFs are analogous to the rest frames of particles. They can be reduced to one non-zero dimension.

---

[1] See Figs. 1.3 and 1.4.

[2] See Fig. 1.5.

[3] The restriction on the vales of r follows from an alternate derivation of spaces based on counting the number of raising and lowering operators in Quantum field expansions for fermions. Odd dimension r spaces have the same number dimensions in their dimension arrays as the next lower even r dimension. The restriction on r values prevents double counting. See our previous books for details.

We found it important to identify each space and dimension array as a fractal.

1.4 Sequence Theory Embodied in Master Equation

The solutions of the Master Equation correspond to various sequences. In the case of our universe, sequences of coupling constants, fundamental fermion masses, strata, and gambol fermions, and so on appear. These are described in chapter 2. They reveal a remarkable ability to unite seemingly disparate quantities. They also provide direct evidence of the features of Cosmos Theory in our universe.

The sequences of Sequence Theory were all found to be fractals in our previous books.

1.5 Parent Universe

There are four components in Cosmos Theory. They are listed in Fig. 1.2. The Master Equation component has a quantum field solution for the Parent universe. We define the Parent universe to be the solution of the Master Equation satisfying a Consistency Condition that balances the virtual force of expansion due to the inner constituents of the Parent universe with the external virtual Casimir force. We found the resulting space-time dimension of this universe to be 18. This universe is the parent of all child universes.[4] (The possibility of several Parent universes is not excluded.)

1.6 Extended Master Equation

Having determined the Master Equation and its solutions we found it significant to generalize the equation to the Extended Master Equation presented in Blaha (2025j). In that book we stated:

> We now introduce a generalization of the Master Equation that has a larger set of coordinates beyond the three Cosmos Space coordinates. Each dimension in the 18 dimension Parent universe dimension array with its 2^{22} dimensions has a corresponding coordinate in the Extended Master Equation. Of these 2^{22} dimensions, we reserve 18 dimensions as space-time dimensions.
>
> ...
>
> In Blaha (2025i) we showed the process for generating the three dimensions of the Cosmos Space from the construction of the Cosmos Fractal Curve. We now similarly generate the 2^{22} new dimensions seen above from HyperSphere (HyperCube) fractal curves for the Parent universe. Later we will apply HyperSphere construction to generate child universe coordinates. *The construction of the new coordinates (dimensions) from 2^{21} fractal curves closely parallels the construction of the two Cosmos Space coordinates from a single fractal – the Cosmos Fractal Curve.*

The form of the Master Equation was extended from 3 dimensions to 3 + 2^{22} dimensions (one dimension for each Parent dimension array component). We found that the additional components shifted the extended equation's energy levels.

---

[4] Higher dimension spaces generate infinities in Perturbation Theory in conventional Quantum Theory. The author's Two Tier formalism for Quantum theory that was developed around 2002 eliminates all such infinities in all dimensions including the fermion triangle infinities. The results of conventional Quantum theory *still hold* in the Two Tier formulation. See Blaha (2002) and (2005a).

As indicated above we can now create the dimensions of the Parent universe as fractal curves. Thus space-time dimensions are constructs. As described in Blaha (2025j) we find:

> We begin by defining a construction process that starts with one dimension line segments as discussed for the Cosmos Fractal Curve construction and grows to generate a 2^{22} dimension unit HyperSphere. This HyperSphere is generated from 2^{21} fractal curves. Each fractal curve generates a *two* dimension unit square which is made continuous and then scaled upwards to a two dimension space. The combined set of two dimension spaces gives a $2 \cdot 2^{21} = 2^{22}$ dimension space. Each dimension of this space corresponds to a dimension in the Parent space and Parent universe dimension array. It is used in the Extended Master Equation. See eq. 4.24.

This development is part of the Cosmos Theory generation process that creates everything – including universes via growth through fractals.

The details of the construction of Cosmos Theory components are described in the following chapters. We begin by showing the major features of particles in our universes follow from Cosmos Theory. The Cosmos directly implies the known, major features of the Unified SuperStandard Theory (UST) and The Standard Model.

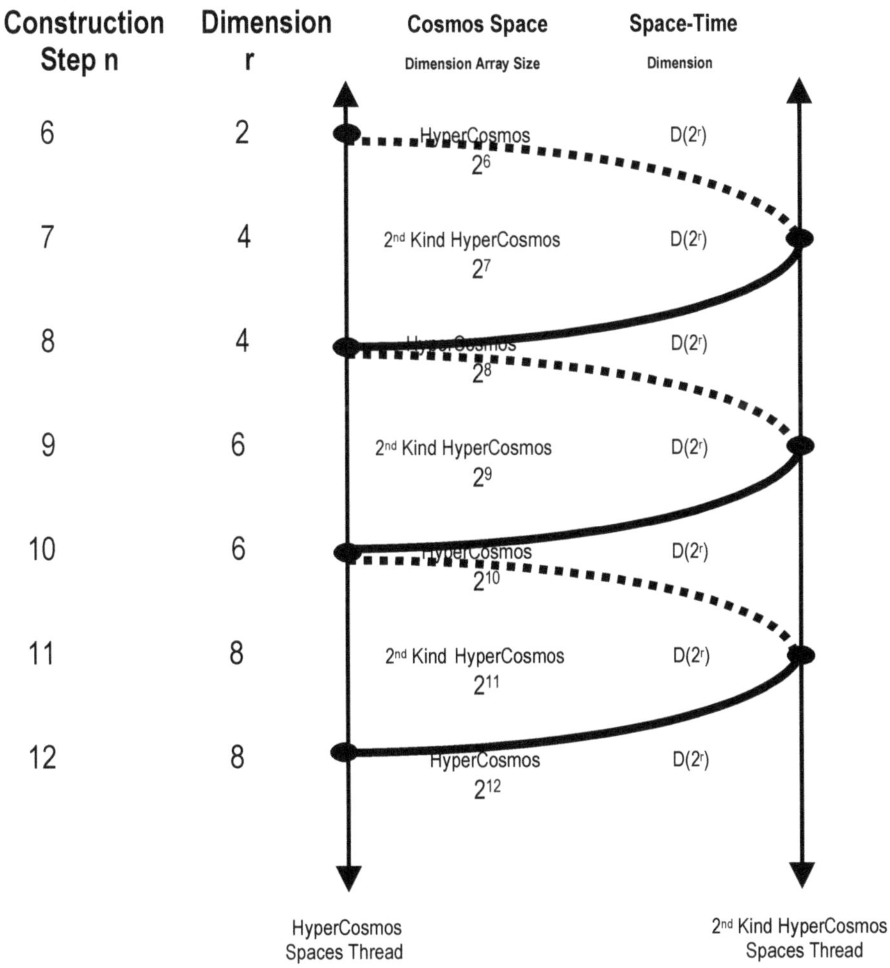

Figure 1.1. A section of the Cosmos Curve showing a flattened double helix form (with the helix mapped to two dimensions.) The two threads of Cosmos spaces and space-times with the "helix" connection between strands are due to the variation in θ. The connecting curved lines form a helix when the 0 spatial dimension of the cutve is considered – an analogue of a double helix similar to DNA. The variation from $\theta = 0$ continuously to $\theta = \pi/(2n)$ takes $D(2^n, \theta)$ from $D(2^{r+4})$ to $D(2^{r+3})$. This change corresponds to a change from n to n + 1 as shown above with solid curving "helix" lines (the "front" view of the helix). The variation from $\theta = \pi/(2n)$ continuously to $\theta = 0$ takes $D(2^n, \theta)$ from $D(2^{r+3})$ to $D(2^{r+4})$. This change corresponds to a change from n to n + 1 as shown above with broken curving "helix" lines (the "rear" view of the helix). The curved lines are not drawn to scale.

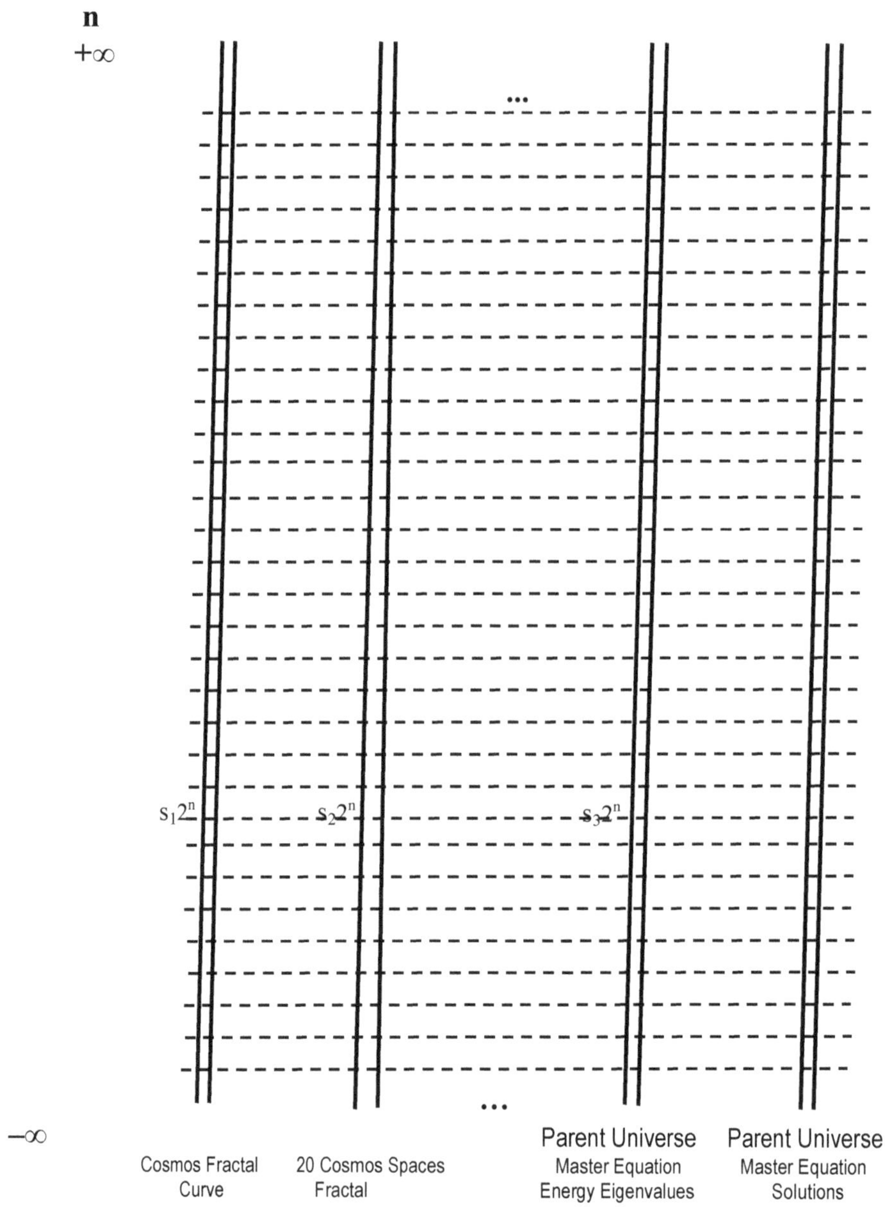

Figure 1.2. The four components of Cosmos Theory. The existence of HyperCosmos and 2nd Kind HyperCosmos spaces causes pairings. The constants s_1, s_2, and s_3 are overall factors for each sequence. The fourth component is the set of Parent universe quantum wave functions $\psi_{n_\varphi n}(u)$ where u is the radial coordinate in Cosmos Space.

THE HYPERCOSMOS SPACES SPECTRUM

Blaha Space Number N_s	Cayley-Dickson Number n_c	Cayley Number Cn d_c	Dimension Array column length $d_{cd} = d_{cr}$	Dimension Array Size $d_{dN}=d_{dr}$	Space-time-Dimension r	CASe Group $su(2^{r/2},2^{r/2})$ CASe
0	10	1024	2048	2048^2	18	su(512,512)
1	9	512	1024	1024^2	16	su(256,256)
2	8	256	512	512^2	14	su(128,128)
3	7	128	256	256^2	12	su(64,64)
4	6	64	128	128^2	10	su(32,32)
5	5	32	64	64^2	8	su(16,16)
6	4	16	32	32^2	6	su(8,8)
7	**3**	**8**	**16**	$\mathbf{16^2}$	**4**	**su(4,4)**
8	2	4	8	8^2	2	su(2,2)
9	1	2	4	4^2	0	su(1,1)
10	0	1	2	2^2	-2	su(0,0)
11	-1	½	1	1	-4	su(-1,-1)

Figure 1.3. The HyperCosmos space spectrum.

THE Limos SECTOR SPACES SPECTRUM

Blaha Space Number $N = O_s$	Cayley-Dickson Number n	Cayley Number d_c	Row Size d_{cr}	Dimension Array Size $d_{dN}=d_{dr}$	Space-time-Dimension r	CASe Group $su(2^{r/2},2^{r/2})$ CASe
10	0	1	2	2^2	-2	U(1)
11	-1	½	1	1^2	-4	U(½)
12	-2	¼	½	$½^2$	-6	U(¼)
13	-3	1/8	¼	$¼^2$	-8	U(1/8)
14	-4	1/16	1/8	$1/8^2$	-10	U(1/16)
				•		
				•		
				•		
∞	−∞	0	0	0	−∞	U(0)

Figure 1.4. The Cosmos Theory Limos spaces spectrum for Gambol Theory. See Blaha (2023e).

HYPERCOSMOS OF THE SECOND KIND SPACES SPECTRUM

Blaha Space Number $N = O_s$	Cayley-Dickson Number n	Cayley Number d_c	Dimension Array size $d_{dN2} = d_{dr2}$	Space-time-Dimension r	CASe Group $su(2^{r/2}, 2^{r/2})$ CASe
0	10	1024	1024 × 2048	18	su(512,512)
1	9	512	512 × 1024	16	su(256,256)
2	8	256	256 × 512	14	su(128,128)
3	7	128	128 × 256	12	su(64,64)
4	6	64	64 × 128	10	su(32,32)
5	5	32	32 × 64	8	su(16,16)
6	4	16	16 × 32	6	su(8,8)
7	**3**	**8**	**8 × 16**	**4**	**su(4,4)**
8	**2**	4	4 × 8	2	su(2,2)
9	**1**	2	2 × 4	0	su(1,1)

Figure 1.5. Part of the HyperCosmos of the Second Kind space spectrum.

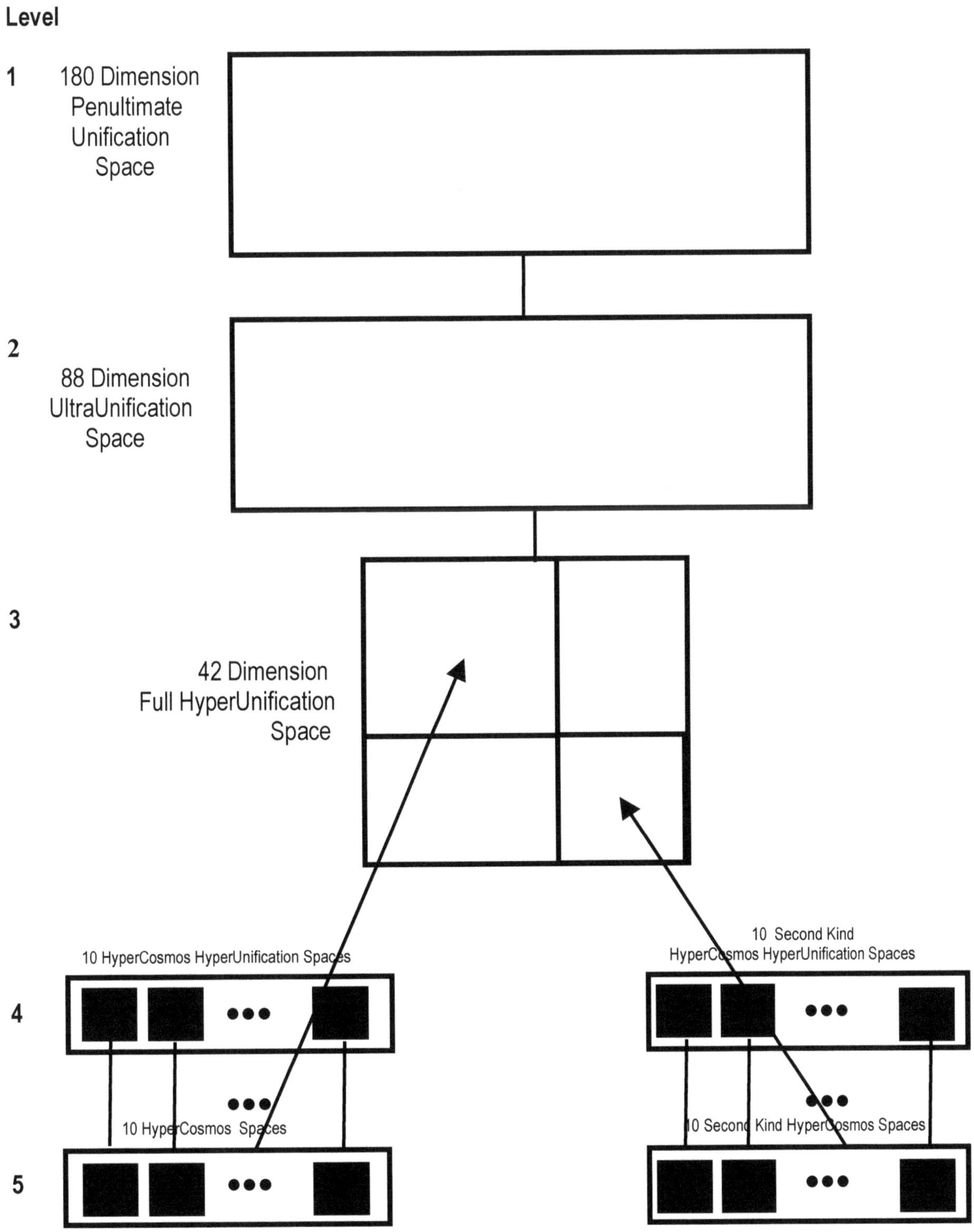

Figure 1.6. Five levels of the Cosmos spaces unification. They contain 43 spaces. From Blaha (2025g).

2. Experimental Evidence for Cosmos Theory

This chapter describes experimentally determined features of elementary particles and their interactions that appear in The Standard Model and in the Unified SuperStandard Theory (UST) *in our universe. They provide accessible evidence for Cosmos Theory.*

2.1 Evidence in the Structure of Fermion Spectrum in Our Universe

The known fundamental quark and lepton fermions are displayed in Fig. 2.1 together with, as yet unfound, fermions of the UST. The fermion spectrum has four layers with each layer consisting of four generations of fermions. (This structuring of fermions into generations differs from the Standard Model structuring but its implications are the same as the results in Standard Model calculations.)

The four generation parts of the layers have an associated U(4) symmetry group that we suggested some years ago. The structuring into layers leads to U(4) Layer groups that we also suggested some years ago in the development of the UST.

The fermion structure of Fig. 2.1, which is for our universe's HyperCosmos space, has both a Normal part of 128 fermions and a Dark part of 128 fermions.

If our universe had a 2nd Kind HyperCosmos space then the Dark part displayed in Figs. 2.1 and 2.2 would be absent. There would only be a Normal part for 128 fermions. The experimental question raised by this difference remains to be resolved.

These fundamental fermion spectrums have been enhanced by the introduction of four strata for each fermion, also the introduction of a corresponding Gambol fermion for each fermion. The resulting generalized structure appears in Fig. 2.3. The factors 95939 and 844242 are determined by Sequence Theory as we will see later. Each fundamental fermion becomes four fermions residing in a separate stratum. Each fermion in each stratum has a corresponding Gambol fermion. Fig. 2.3 shows the expansion of one fundamental fermion into eight fermions. This expansion takes place for all 256 HyperCosmos fermions in our universe resulting in 2048 distinct fermions if Dark fermions exist. For a 2nd Kind HyperCosmos space and universe this expansion takes place for all its 128 fermions resulting in 1024 distinct Normal fermions.

In general, a (HyperCosmos space) universe of dimension r has an expanded dimension array and corresponding expanded fermion spectrum of size 2^{r+7}. A 2nd Kind HyperCosmos space universe of dimension r has an expanded dimension array and corresponding expanded fermion spectrum of size 2^{r+6}

The Fermion Periodic Table – Our Universe[5]

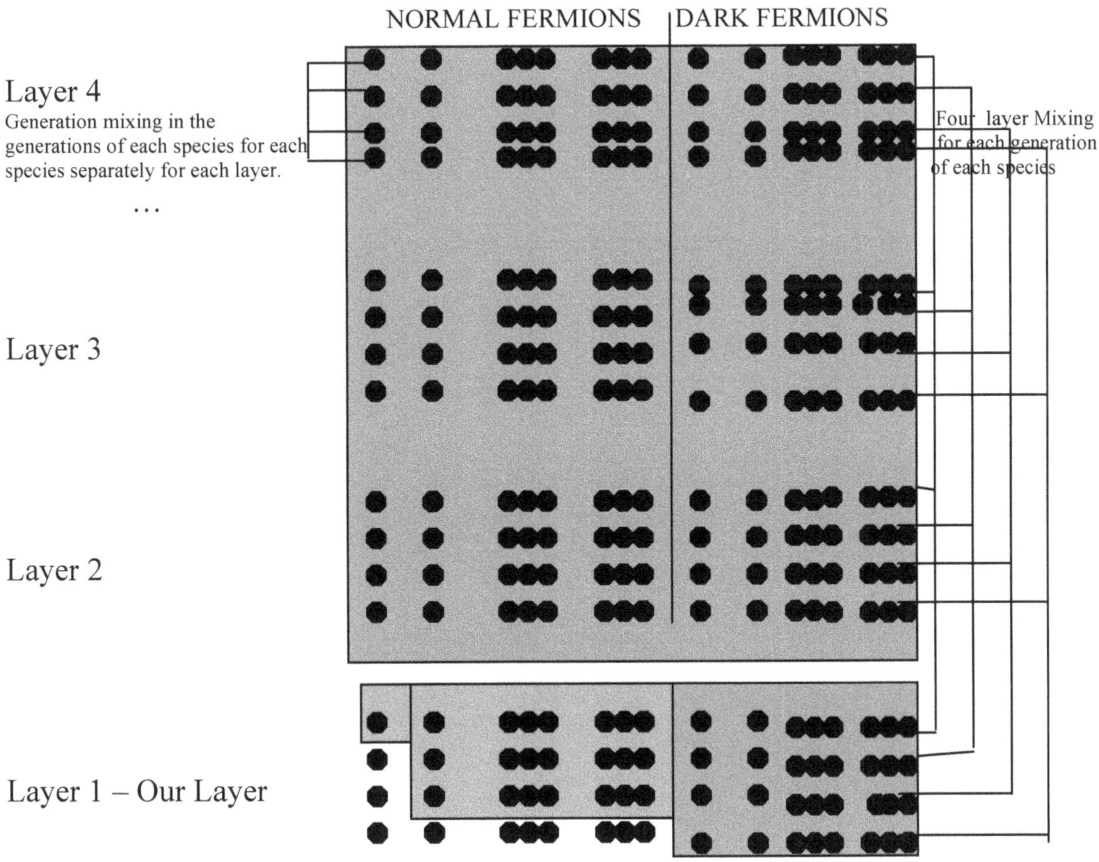

Figure 2.1. The UST fermions show nested partitioning based on the *SU(n) dimension arrays*. The Fermion particle spectrum and partial examples of the pattern of mass mixing of the Generation groups and of the Layer groups. Unshaded parts are the known fermions. One shaded generation is an additional, as yet not found, 4th generation. The lines on the left side (only shown for one layer) display the Generation Group mixing within each layer. The Generation mixing occurs within each layer using a separate Generation group for each layer. The lines on the right side show Layer group mixing (for Dark matter) with the mixing among all four layers for each of the four generations individually. There are four Layer groups for Normal matter and four Layer groups for Dark matter. For a HyperCosmos space universe there are 256 fundamental fermions.

---

[5] From Blaha (2025g) and earlier books.

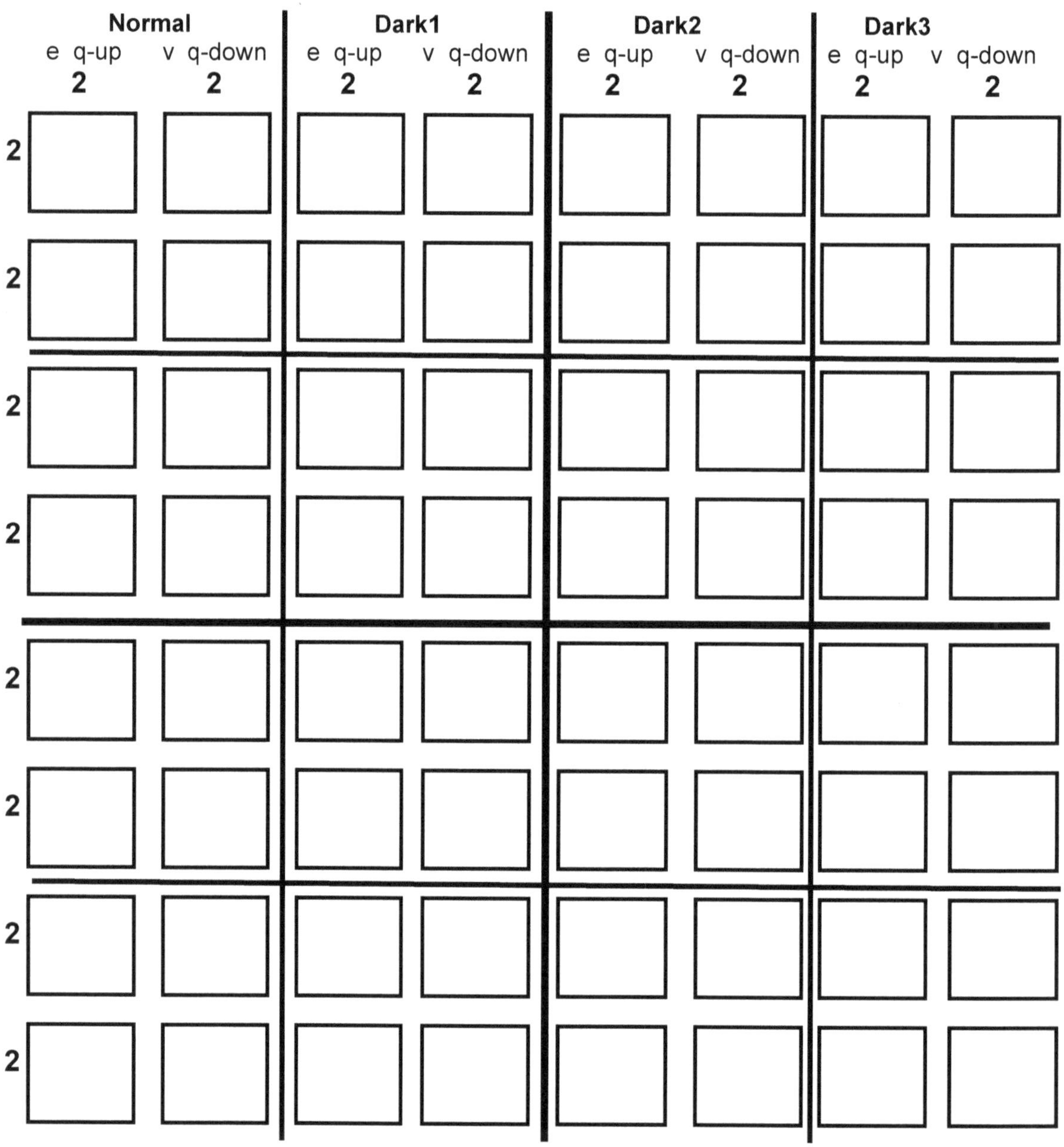

Figure 2.2. Partition of the fermion 16 × 16 dimension array of our universe (if it has a HyperCosmos space): 4 layers of 4 generations. The label e q-up indicates a charged lepton – up-type quark pair. The label v q-down indicates a neutral lepton and a down-type quark, and so on. The Megaverse of dimension 6 would have a 32 × 32 dimension array and a corresponding set of fundamental fermions.

Strata and Gambol Fermions[6]

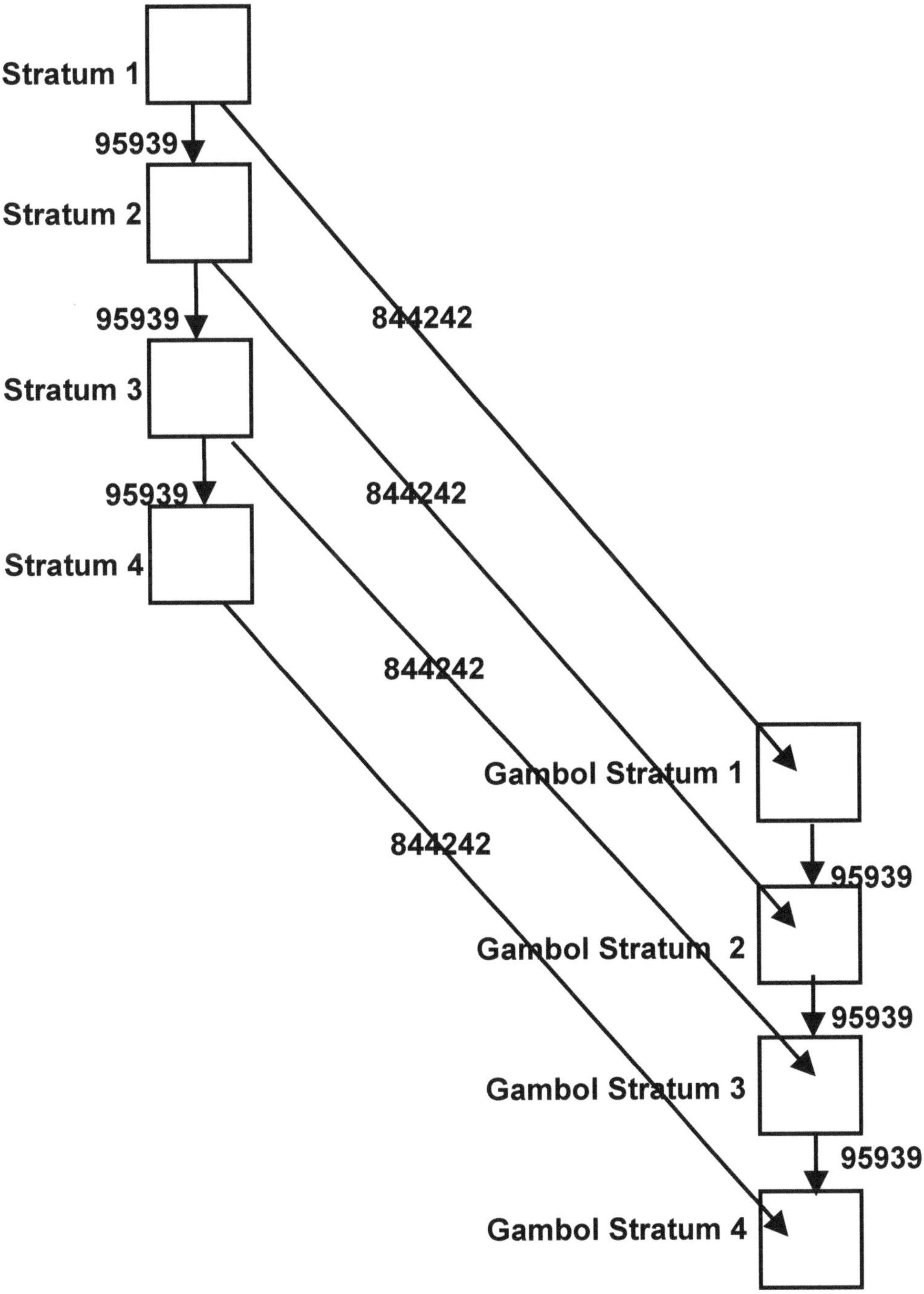

Figure 2.3. Each fermion in each stratum has a corresponding Gambol fermion. One fundamental fermion splits into eight fermions. This expansion takes place for all 256 HyperCosmos fermions in our universe resulting in 2048 distinct fermions if Dark fermions exist. Coresponding fermions in adjacent strata differ in mass by a factor of 95939

---

[6] From Blaha (2025g) and earlier books.

in increasing order from stratum 1. Each gambol fermion is a factor of 1/844224 of the corresponding fermion mass. Each gambol mass is 1/95939 of the mass of the gambol immediately above it. Not drawn to scale. All fundamental fermion particles pictured above are fractal.

2.1.1 Fractal Nature of Particles

Elementary particles are created from a form of point interaction in Quantum Field Theory. A comparison of the nature of this creation process with the generation of Cosmos spaces and universes, and of the Cosmos itself, all of which are fractals grown from mathematical points, suggests that elementary particles have a fractal nature as well. We have explored this idea in our previous books.

2.2 Evidence in the Structure of Symmetries in Our Universe

The symmetries found in the Elementary Particle Theories of our universe have a pattern that embodies the structure of our universe's dimension array. The pattern of symmetries mirrors that of the pattern of fermions in Fig. 2.2. Both types of patterns are determined by a map from our universe's dimension array. Fig. 2.4 displays the pattern of symmetries in our universe if it has a HyperCosmos space.

Thus we find the fundamental fermions and the symmetries of our universe can be viewed as maps from our universe's dimension array – credible evidence in favor of Cosmos Theory.

Internal Symmetries – Our Universe[7]

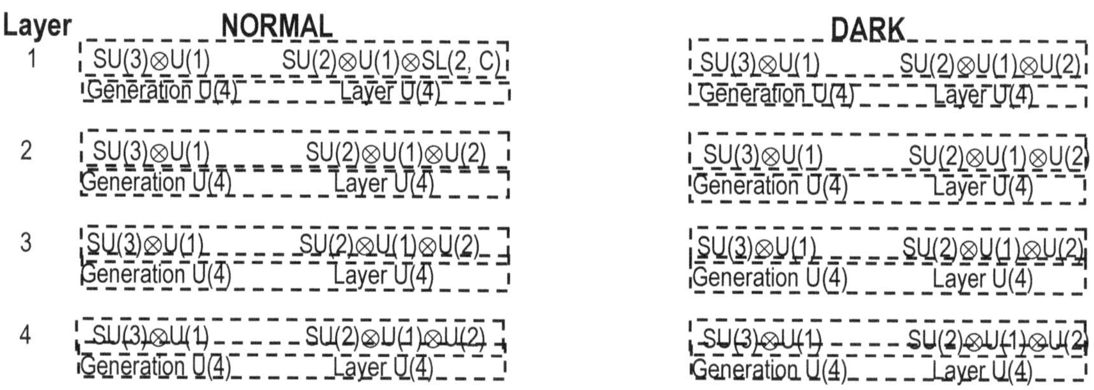

Figure 2.4. The sets of symmetries in the UST for our universe if it has a HyperCosmos space. Note each boxed element above has a pair of 8 real dimension irreducible representations. The SL(2, **C**) representation has four real dimensions.[8] If the Dark part is removed it gives the UST symmetries for a universe of a 2nd Kind HyperCosmos space.

---

[7] From Blaha (2025g) and earlier books.

[8] The Lorentz Group $SO^+(1, 3)$ is often specified with an SL(2, **C**) representation.

2.3 Evidence in the Experimental Implications of Cosmos in Sequence Theory

The consideration of the solutions of the Master Equation[9] (and the Extended Master Equation) leads to sets of solutions parameterized by the quantum number n_φ. The basic equations are

$$\backslash[(\partial^2/\partial x^2 + \partial^2/\partial y^2)/m + \lambda/m^2r^2 + C)\psi = 0$$
$$(\partial^2/\partial u^2 + 1/u\ \partial/\partial u + \lambda_{n\varphi}/u^2 + mC)\ \psi(u) = 0 \qquad (2.1)$$

where[10]

$$\lambda_{n\varphi} = n_\varphi{}^2/m \qquad (2.2)$$

and m = 0.7696. They give an energy squared:

$$C_{n\varphi n} = c_{n\varphi} \exp(n\pi/(\lambda_{n\varphi} - ¼)^{½}) \qquad (2,3)$$

where n is the radial quantum number and n_φ is the angular momentum quantum number. The growth factor (multiplier) for each power series sequence is

$$g_{n\varphi} = \exp(\pi/(\lambda_{n\varphi} - ¼)^{½}) \qquad (2.4)$$

The relevant choices of the values of n_φ are listed in Fig. 2.5.

The evidence for Cosmos Theory that emerges from its Sequence Theory subsector has two forms: the individual evidence of each of the entries in the Sequence Table of Fig. 2.5, and the combined evidence for Sequences Theory as generated from the Master Equation for the series of n_φ values in the table. The n_φ values are sequential powers of 2 for rows A1 – C and E. This conforms to the view that n_φ is a quantum number analogous to angular momentum *l* in the case of the Hydrogen atom model. The other values of n_φ in rows D and F are for fermion mass sequences. They also display a similarity being related to the square root of two, $2^{½}$.

For these reasons the Sequences Table also is evidence in our universe for Cosmos Theory.

Note: we take the $g_{n\varphi}$ and n_φ values to be fractal numbers. See our previous books such as Blaha (2024b).

2.3.1 Direct Evidence for the Sequence Theory Part of Cosmos Theory

The sequences in rows of A1 – G of Fig. 2.5 are directly generated from the Master Equation. Each value of $\lambda_{n\varphi}$ (and thus n_φ) generates a sequence of solutions for $C_{n\varphi n}$ that becomes values for a sequence of quantities $C_{n\varphi n}$. The factor $g_{n\varphi}$ is the growth factor (sequence values multiplier) for the respective sequence. The entries in Fig. 2.5 give coupling constants and fermion masses. These sequences, plus Higgs masses and the masses of as yet unknown particles are the known content of The Standard Model

[9] Blaha (2025c) and (2025f).

[10] The eigenvalue spectrum results from the imposition of orthogonality condition on the solution function near u = 0 according to von Neumann.

and UST. Particle scattering amplitudes, and crossections, and certain quantities such as magnetic moments remain to be found. Thus our sequences are close to a complete description of the constants of the Standard Model and UST. The, as yet currently unkown items, have both theoretical problems in their calculation and experimental problems in going to the required higher energies.

It is somewhat amazing to see so much of the UST and Standard Model generated directly from the sequences derived from the Master Equation.

Note the spaces, and unification spaces, of Cosmos Theory are also determined directly as well. Thus there is a unity in the formulation of this disparate set of quantities.

SEQUENCES TABLE[11]

Sequence Fractal	$\mathbf{n_\varphi}$	g_{n_φ} **Fractal Dimension multiplier factor**	$\lambda_{n\varphi} = n_\varphi^2/m$	$\lambda_{n\varphi} = \pi^2/(\ln g_{n\varphi})^2 + ¼$ **Calculated**
Fractal	12	$1{,}26 = \ln_3(4)$	187.1	185
Fractal	10	$1.317 = \ln_{2.7}(3.7) = \ln_5(8)$	129.9	130.4
Fractal	8	$1.41 = \ln_{2.36}(3.36) = \ln_4(7)$	83.16	83.9
Sierpinski Triangles Fractal; 1.58-bit Large Language Model	6	$1.585 = \ln_2(3)$	46.778	46.77
A1 Cosmos Fractal Curve line steps	4	$2 = \ln_2(4)$	20.79	20.79
A2 Cosmos spaces of dimension r dimension array column lengths $2^{r/2+2}$	4	$2 = \ln_2(4)$	20.79	20.79
A3 The sequence of Unification Spaces s	4	$2 = \ln_2(4)$	20.79	20.79
B Coupling Constants α Multiplier	2	$4 = \ln_2(16)$	5.398	5.39
C Fermion Sequence 2 Masses	1	$2/3 \times 2^5 = \ln_2((32^{32})^{0.133})$	1.3	1.3
D Fermion Sequence 1 Masses	$½^{½}$	$3/2 \times \pi 2^5 = \ln_2((32^{32})^{0.3\pi})$	0.65	0.81
E Fermion Strata Masses Sequence	½	$95939 = 5.96 \ln_2(32^{32\cdot 32\pi})$	0.325	0.325
F Fermion Gsmbol masses factors based on α_{QED}	$0.48 \approx 2^{½}/3$	$844242 = 5.15 \ln_2(32^{32\cdot 32\cdot 32})$	0.299	0.303
G n_φ Values Sequence; A Sequence of Sequences	2	$4 = \ln_2(16)$	5.398	5.39

Figure 2.5. Extended Cosmos Theory Sequences with sequence F determined by setting $\lambda = 0.303$, which is the value of the QED couping constant e that determines the Fine Structure Constant α. The first row is for the case of a multiplier √2. Its n_φ value is an extension of the series of n_φ values 2, 4, 8 of rows A1 through B. The first and second rows are for the cases $n_\varphi = 6$ and $n_\varphi = 8$ which also may be of interest. The second row may correspond to 1.58-bit Large Language Models. Note: $32\cdot 32 = 32^2$ and $32\cdot 32\cdot 32 = 32^3$ in items D, E, and F above. Note the "Calculated $\lambda_{n\varphi}$ " values are determined from the $g_{n\varphi}$ values by $\lambda_{n\varphi} = \pi^2/(\ln g_{n\varphi})^2 + ¼$.

2.3.1 Explanation of Sequences

We now consider the meaning of some sequences in Fig. 2.5. First note the first line A applies to the structure of Cosmos Theory in general. Most of the remaining lines are for individual universes. We view these lines as applicable in all universes of all ten Cosmos spaces. The features of the individual sequences in the above rows are:

A Cosmos dimension array column sizes ($2^{r/2+2}$) Multiplier $\quad n_\varphi = 4$

[11] From Blaha (2025i) and earlier books such as (2025a) and (2025e).

Cosmos spaces' dimension array columns and rows each grow by factors of 2 as the dimension increases by a factor of 2. The dimension array size grows by factors of 4. This sequence directly matches the sequence of line segments in the Hilbert fractal curve construction as well as the sequence of line segments in the fractal Cosmos Curve construction.

B Coupling Constants Multiplier $n_\varphi = 2$
The coupling constants sequence grows by factors of 4. See below detail in section 2.4.

C Fermion Sequence 2 Masses $n_\varphi = 1$
The sequence 2 fermions grow by factors of 32. See section 2.5.

D Fermion Sequence 1 Masses $n_\varphi = ½^{½} = 2^{½}/2 = 0.707107$
The sequence 1 fermions grow by factors of 32π. See section 2.5. This n_φ value has a multiplier of $2^{½}$ from the $n_\varphi = 1$ value and also to the $n_\varphi = ½$ value. Thus it is mid-sequence. It has the value ½ times $2^{½}$ - multiplicatively greater than the sequence value of the next $n_\varphi = ½$ value.

E Sequence of n_φ Values $n_\varphi = 1/2$
This is the sequence of n_φ values for fermion strata. The items in line E are:

$$\lambda_{n_\varphi=½} = n_\varphi^2/m = ¼/m = 0.325$$
$$g_{g½} = \exp(\pi/(\lambda_{n_\varphi=½} - ¼)^{½})$$
$$= 95939$$

The sequence of $n_\varphi = ½$ values has a growth multiplier $g_{g½} = 95939$. This extremely large value is, at first glance, somewhat strange. It is much larger than the analogous multipliers for fermions in items C and D. However, the relation of the ν' and ν_e Neutrino masses seen later suggests that there may be some new parts to the complete spectrum of fundamental fermions. We see[12]

$$\nu' = \nu_e\, e^2(\alpha_0^2)^{-1} \approx \pi e^2 g_{g½}\nu_e$$

where

$$g_{g½} = 95939 \approx 1/(\pi\alpha_0^2) = 97852$$

as in row E above using

$$\alpha_0 = e^2/4096 = 0.0018036$$

Thus we may extrapolate and find a sequence of neutrino masses based on the mass multiplier of 95939. (The C and D items are also for sequences of fundamental

---

[12] The choice of the ν' mass was made in our earlier book. It was based on making sequence 2 masses approximately follow a power sequence form, which was not possible if we used the ν_e mass.

fermions.) It is then reasonable to assume that all fundamental fermions in our universe are "multi-plicated" into fermion strata.

We define a number n_s labeling a proposed set of replicates, strata, of the 256 fermions of the r = 4 Cosmos universe of UST. One replicate of 256 fundamental fermions that includes ν_e can be viewed as the $n_s = 1$ set. The next $n_s = 2$ set has the ν' neutrino and 255 additional fermions with each of their masses scaled up from their $n_s =$ 1 counterpart by a factor of $\pi e^2 g_{g^{1/2}}$.

Seeing that there are four sequences above and that they have a lower limit of the $n_\varphi = ½$ sequence and an upper limit conceptually[13] of the $n_\varphi = 4$ sequence we suggest that there are 4 counterparts, strata, of the 256 fermion spectrum in our universe with corresponding results in other universes of other spaces. We call each counterpart a *stratum* and label them as $n_s = 1, 2, 3, 4$. See Fig. 2.3.

The set of strata supports an extension of the fermion Cosmos Theory dimension array. We use them to define a Sequence Theory fermion dimension array of 2^{r+6} dimensions due to a factor of 2^2 from the four strata. We do not introduce new vector or scalar boson strata.

F Fermion Gsmbol masses Sequence based on α_{QED} $0.48 \approx 2^{1/2}/3$ $844242 = 5.15 \ln_2(32^{32 \cdot 32 \cdot 32})$ 0.299 0.303

This row is based on choosing the λ_{n_φ} value[14] equal to the Fine Structure Constant $g = (4\pi\alpha_{QED})^{1/2}$. The quantity λ_{n_φ} is a coupling constant. See Fig. 2.6. We use the number 844242 as the factor giving the value of the Gambol Fermion mass obtained from the corresponding fermion mass M: Gambol Mass = M/844242.

G n_φ Values Sequence: A Sequence of Sequences 2 $4 = \ln_2(16)$ 5.398 5.39

This row gives the factor of 2 seen in a subset of n_φ values seen in Fig. 2.5 in rows A1 – A3, B, C, and E.

2.4 Direct Evidence in the Coupling Constant Sequence

There are three known coupling constants in the Standard Model and the UST. They are listed as column g in Fig. 2.6. They form a sequence in g as the group of the interaction increases. Row by row they double. Their coupling $\alpha_n = g_n^2/4\pi$ grows by factors of four. Thus they correspond to row B in Fig. 2.5. The form of α_n is

$$\alpha_n = g_n^2/4\pi = \alpha_0 2^{2n} \quad (2.5)$$

---

[13] Partitioning dimension arrays further is not significant. Also the g_g value is close to 1 making it of less interest.

[14] Later we show the $\lambda_{n\varphi}$ values are coupling constants too.

Coupling Constants – Our Universe[15]

	ESTIMATE			EXPERIMENT		
Interaction	**Expression**	$\alpha_i = g^2/4\pi$ **Value**	**g**	**Known[16] Value** $g^2/4\pi$	**Deviation**
U(0) α_0	$e^2/4096$	0.0018036	-	-	-
U(1) $\alpha_1 = \alpha$	$e^2/1024$	0.00721438	0.303	0.0072973525643	1.15%
SU(2) $\alpha_2 = g^2/4\pi$	$e^2/256$	0.0289	0.63	0.0316	9.3%
SU(3) $\alpha_3 = \alpha_S$	$e^2/64$	0.115	1.21	0.117	1.7%
SU(4)[17] α_4	$e^2/16$	0.462	2.4?	0.458	0.087%

Figure 2.6. Coupling constants for the UST (and the Standard Model). Note factor of four relation between adjoining α values.[18]

2.4.1 Direct Relation between Coupling Constants and Symmetry Groups

The known coupling constants are accurately derived from the row B sequence of Fig. 2.5. They form a sequence based on powers of 2 that follows from the scaling feature of the Master Equation solutions for $n_\varphi = 2$. See Fig. 2.6.

The group numbers n of the corresponding symmetry groups SU(n) also form a sequence. This sequence has equally spaced values (and is thus *additive*) in the parameter n of SU(n): n = 1, 2, 3, … Therefore, it initially does not appear to be generated by the Master Equation. However if the Master Equation is transformed into an approximate harmonic oscillator form its eigenvalues are equally spaced.

In Blaha (2025f) we showed that the scaling Master Equation solutions can be related to harmonic oscillator solutions at short distances. Thus we can view the group number n as a harmonic oscillator eigenvalue.

Comparing the scaling sequence with the equally spaced sequence we obtain a relation between group number and its corresponding coupling constant value. *Thus we have a new view of the features of Yang-Mills gauge fields relating group number and coupling constant value.*

Proof of Relation Between Master Equation Coupling Constants

In Blaha (2025f) we found the quantized energy $E_{vn\varphi n}$ of the approximate harmonic oscillator form of the Master Equation:

$$E_{vn\varphi n'} = n'\pi/(\lambda_{n\varphi} - 1/4)^{1/2} + \text{const} \quad (2.6)$$

where n' is the eigenvalue, $\lambda_{n\varphi}$ appears in columns of Fig.2.5, and "const" shifts the zero point of the oscillator. Eq. 2.6 agrees with the scaling form of the Master Equation quantization that gives the coupling constant value:

---

[15] From Blaha (2025g) and earlier books.
[16] All coupling constant values are based on data from Particle Data Group Tables of 2024.
[17] This value is based on the "doubling trend" seen in the three known coupling constants g_i above in Fig. 3.2.
[18] From Blaha (2018e).

$$C_{n\varphi n} = c_{n_\varphi} \exp(2n\pi/(\lambda_{n_\varphi} - ¼)^{½}) = c_{n_\varphi} \exp(2E_{vn\varphi n}) \quad (2.7)$$
$$C_{n\varphi n}/c_{n_\varphi} = \alpha_0\, 2^{2n}$$

We identify n' as the parameter of SU(n'). We can calculate n' from eq. 2.6:

$$n' = 2(E_{vn\varphi n} - \text{const})\, (\lambda_{n_\varphi} - ¼)^{½}/\pi \quad (2.8)$$
$$= 2(\, ½\ln(C_{n\varphi n}/c_{n_\varphi}) - \text{const})\, (\lambda_{n_\varphi} - ¼)^{½}/\pi$$

Using

$$C_{n\varphi n}/c_{n_\varphi} = \alpha_0\, 2^{2n} \quad (2.9)$$

then from eq. 2.5 we find

$$n' = 2(½(\ln(\alpha_0) + 2n\ln(2)) - \text{const})\, (\lambda_{n_\varphi} - ¼)^{½}/\pi \quad (2.10)$$
$$= 2n\ln(2)\, (\lambda_{n_\varphi} - ¼)^{½}/\pi + (\ln(\alpha_0) - 2\,\text{const})\, (\lambda_{n_\varphi} - ¼)^{½}/\pi$$

If we set the "const" value:

$$\text{const} = ½ \ln(\alpha_0) = \ln(\alpha_0^{½}) \quad (2.11)$$

and note

$$2\ln(2)\, (\lambda_{n_\varphi} - ¼)^{½}/\pi = 1$$

if $\lambda_{n_\varphi} \cong 5.39$ as it does in the B column of Fig. 2.5 for $n_\varphi = 2$, then

$$n' = n \quad (2.12)$$

Thus the arithmetic n' for the approximate harmonic oscillator Master Equation solution equals the multiplicative n of the scaling form of the Master Equation. This result is independent of the dimension r of the space.

For all Cosmos spaces the SU(n) parameter n determines the coupling constant value. *Coupling constants form a universal set of values.* Should life exist in other universes their interactions will have the same coupling constants, and thus life there is likely similar to life in our universe.

The intimate connection between symmetry groups and their coupling constants is an experimental vote of confidence for Cosmos Theory and its Master Equation.

2.4.2 Yang-Mills Field Equations Determined

The Yang-Mills gauge fields had their Lagrangian form determined by their Yang-Mills symmetry. The values of their coupling constants were considered arbitrary.

Our above derivation of coupling constant values, g, (in Fig. 2.6) agrees with experiment for our universe, and we now see applies to *any* universe. Consequently, the values of Yang-Mills, gauge field, coupling constants are now also fully determined.

The form of the Yang-Mills gauge field expressions are now also completely determined for each group:

$$F^{\mu\nu} = \partial^\mu A^\nu - \partial^\nu A^\mu + ig[A^\mu, A^\nu] \quad (2.13)$$

and the Lagrangian is

$$\mathscr{L} = -\tfrac{1}{4}\, F^{\mu\nu}\cdot F_{\mu\nu} \tag{2.14}$$

At this point we now have a complete characterization of the gauge theory sector of The Standard Model and the Unified SuperStandard Theory (UST).

2.4.3 Cosmos Theory Calculation of the Weinberg Angle Value

The Weinberg angle is well approximated by the Fig. 2.6 values of the SU(2) and U(1) coupling constants:

$$\tan(\theta_W) = 0.303/0.63 = 0.48$$
$$\sin^2(\theta_W) = 0.19$$

which is quite close to known approximate experimental value (according to CODATA 2022):

$$\sin^2(\theta_{W\text{experiment}}) = 0.22\ (23)$$

Thus we find additional support for the Coupling Constant sequence that follows from the Master Equation determination of coupling constants.

2.4.4 Cosmos Theory Calculations and Renormalization

In Quantum Field Theory physically interesting quantities such as coupling constants have an initial value usually determined at accessible lower energies. They are then subject to renormalization such as we have shown in our calculation of the Fine Structure Constant α in massless QED. Are there renormalization effects in the structure of Cosmos Theory? No, Sequence Theory sequences are based on the Master Equation without interactions.[19] Therefore the Master Equation has no renormalization.

The sequence table of Coupling Constants (Fig. 2.6) shows a simple progression in coupling constants between the various symmetry groups that agrees with experiment. Are there renormalization effects implicit in these coupling constants? It is possible that these effects appear in the starting coupling constant value α_0. Their appearance in the other constants in the sequence is muted by the regularity of the sequence. One can only view the regularity as a constraint on the renormalization of the constants that forces the regularity, which is determined by the Master Equation. Thus the regularity is consistent with any implicit renormalization effects. The calculation of renormalization effects on coupling constants must be done "individually." We have discussed the adaptation of our α calculation to other coupling constant calculations in chapters 1 and 2 of Blaha (2025b).

The connection of the Cosmos sequences to elementary particle experimental features in our universe (and in other universes for coupling constants) is remarkable and can only be viewed as significantly constraining renormalizations. We return to this topic later.

[1919] Also the Master Equation and Sequence Theory do not exist in a space-time. Thus no Perturbation Theory applies.

The remaining great issue of particle theory is the source and values of particle masses, to which we now turn. The determination of scattering amplitudes and other constants such as magnetic moments also require elucidation.

2.5 Cosmos Theory Confirming Evidence from Fermion Masses Sequences

At first glance there are no obvious patterns in the masses of the known fundamental fermions. Fig. 2.7 displays their experimental values and some simple revisions suggested by the form of the data.

In previous books we first developed an approximate pair of patterns for subsets of the set of fermions. These patterns are approximately accurate. Deviations from the patterns appear to be due to additional contributions from interactions, which we describe later.

α/β	REVISED ESTIMATE	DEVIATION RATIO	DATA
1	$\nu_e = 0.0372$ ev/c^2 $= 3.72 \times 10^{-11}$ GeV/c^2		
1	$\nu' = \nu_e\,(\alpha_0{}^2)^{-1} = 8.4 \times 10^{-5}$ GeV/c$^2 \cong e2^{-15}$ GeV/c^2		
1	$e = r_1\,\nu' = 2\,\pi\,\nu' = 0.511 \times 10^{-3}$	1.	0.511×10^{-3} GeV/c^2
1	$d = r_2 e = r_2 r_1 \nu' = 16\pi\,\nu' = 4.24 \times 10^{-3}$	1.13	3.76×10^{-3} GeV/c^2
¾	$u = r_3 d = r_3 r_2 r_1 \nu' = ¾ \times 2^3 \pi\,\nu' = 1.80 \times 10^{-3}$	1.02	1.76×10^{-3} GeV/c^2
¾	$c = r_4 u = r_4 r_3 r_2 r_1 \nu' = ¾ \times 2^{11} \pi^2\,\nu' = 1.28$	1.001	1.27 GeV/c^2
¾	$s = r_5 c = r_5 r_4 r_3 r_2 r_1 \nu' = ¾ \times 2^9 \pi\,\nu' = 102 \times 10^{-3}$	1.07	95×10^{-3} GeV/c^2
½	$t = r_6 s = r_6 r_5 r_4 r_3 r_2 r_1 \nu' = 2^{16} \pi^3\,\nu' = 171$	0.99	172.76 GeV/c^2
½	$b = r_7 t = r_7 r_6 r_5 r_4 r_3 r_2 r_1 \nu' = 2^{14} \pi\,\nu' = 4.34$	1.04	4.18 GeV/c^2

Figure 2.7. Revised fermion masses based on α/β and on the known value of e of 0.511 from which the values of ν' and ν are extracted and used to calculate the other fermion masses. The ν' mass (for a hypothetical neutrino) was intriduced to establish an entry in the pattern denoted Sequence 2 seen later. The d discrepancy is the largest reflecting, perhaps, some unceratinty in the experiment value of d. Otherwise there is excellent agreement with experiment!

Our reconstruction of the data is based on the possibility of improved mass estimates by the introduction of a factor giving a sliding scale of mass dependence on the ratio of an ElectroWeak SU(2) factor α and a Strong Interaction SU(3) factor β. We assume they contribute with equal strength to the e and d masses; with a mild dominance of ElectroWeak over Strong for u, c, and s quarks; and with dominance of Strong over ElectroWeak by a factor of 2 for t and b quarks.

We denote the ratio of the interaction strengths with

$$\alpha/\beta \equiv \alpha_{SU(2)}(m)/\beta_{SU(3)}(m) \qquad (2.15)$$

where m is the mass of the relevant fermion. We note that the ratio of 2 for t and b reflects the ratio of the coupling constants (1.21/0.63 = 1.92 ≅ 2) in Fig. 2.7. The value of α/β nicely progresses from 1 to ¾ to ½ as the fermion mass increases.

Strong SU(3) dominates at high mass; ElectroWeak SU(2) dominates at low mass. All estimates in Fig. 2.7 are close to their experimental values. We also have a representation of fermion masses in terms of powers of 2, powers of π, and the value of v' - comparable to coupling constants where coupling constants are powers of 2 multiplied by e – the base of natural logarithms. This suggests that Cosmos Theory is at the base of both coupling constant and fermion mass values. Note: one uses π and e, and the other uses only e - the base of natural logarithms.

Substituting $v' \cong e2^{-15}$ GeV/c^2 gives Fig. 2.8.

α/β	**FERMION**
1	$\nu_e = 0.0372$ ev/c^2 $= 3.72 \times 10^{-11}$ GeV/c^2
1	$\nu_e = \alpha_0^2\, v' = e^5 2^{-39}$ GeV/c^2
1	$v' = \nu_e\, (\alpha_0^2)^{-1} \cong e2^{-15}$ GeV/c^2
1	$e = r_1\, v' = \pi e 2^{-14}$ GeV/c^2
1	$d = r_2 e = r_2 r_1 v' = \pi e 2^{-11}$ GeV/c^2
¾	$u = r_3 d = r_3 r_2 r_1 v' = ¾ \times \pi e 2^{-12}$ GeV/c^2
¾	$c = r_4 u = r_4 r_3 r_2 r_1 v' = ¾ \times \pi^2 e 2^{-4}$ GeV/c^2
¾	$s = r_5 c = r_5 r_4 r_3 r_2 r_1 v' = ¾ \times \pi e 2^{-6}$ GeV/c^2
½	$t = r_6 s = r_6 r_5 r_4 r_3 r_2 r_1 v' = 2\pi^3 e$ GeV/c^2
½	$b = r_7 t = r_7 r_6 r_5 r_4 r_3 r_2 r_1 v' = \pi e 2^{-1}$ GeV/c^2

Figure 2.8. Revised form of fermion masses using $v' \cong e2^{-15}$ GeV/c^2. The first two rows are equal expressions for ν_e.

An examination of the data in Figs. 2.7 and 2.8 suggested two sequences of fermions.[20] One sequence is for "up-type" fermions and the other for "down-type" fermions. The approximate masses for the two fundamental fermion sequences appear in Fig. 2.9.

[20] See Blaha (2024g) and (2024h) for a detailed derivation of these mass sequences.

Fermion Sequences – Our Universe[21]

	SEQUENCE 1				SEQUENCE 2			
	e	u	c	t	ν'	d	s	b
Mass:	0.511×10^{-3}	1.80×10^{-3}	1.28	171	8.4×10^{-5}	4.24×10^{-3}	102×10^{-3}	4.34
Multiplier:	$2^5\pi$	$2^5\pi$	$2^5\pi$		32	32	32	

Figure 2.9. The fundamental fermion mass sequences with approximate multiplier factors between the masses of each sequence. The sequence 1 masses show a significant variation with their $2^5\pi$ factor but their overall trend is approximately satisfactory. Sequence 2 masses better approximate their factor of 32 progression. These sequences are analogous to the coupling constant sequence seen above. *Later we show a direct relation to the coupling constant sequence symmetries.*

We suggest all four generations of fermions implement this separation into sequences. The two sequences implement the multipliers:

Up type fermion sequence multiplier: $2^5\pi = 32\pi$
Down type fermion sequence multiplier: $2^5 = 32$

We believe the multiplier of $2^5 = 32$ appearing in both sequences is the number of fermions in a UST layer based on the concept that all fermions of a layer are intertwined. There are 32 normal fermions in a UST layer. The explanation of the factors of π may reflect gambol effects.

2.5.1 Cosmos Theory Mass Calculations and Renormalization

In Quantum Field Theory physically interesting quantities have an initial value usually determined at accessible lower energies. They are then subject to renormalization such as we have shown in our calculation of the Fine Structure Constant α in massless QED. Cosmos Theory parts are based on the Master Equation without interactions. Therefore there are no renormalization effects on its predictions.

The above fermion mass sequences (Fig. 2.9) show simple progressions in masses that approximately agree with experiment. Are there renormalization effects implicit in these progressions? Most likely yes. They appear to generate deviations from the progression values for some masses. The renormalization effects appear muted by the regularity of the sequences. One can only view the approximate regularity as a consequence of the basic strength of the Cosmos sequences that are determined by the Master Equation.

The connection of the Cosmos sequences to elementary particle experimental features in our universe (and in other universes for coupling constants) is remarkable and can only be viewed as providing a fundamental underpinning of renormalizations. We return to this topic in chapter 3.

---

[21] From Blaha (2025g) and earlier books.

The remaining great issue of particle theory is the source and values of particle masses, to which we now turn. The determination of scattering amplitudes and other constants such as magnetic moments also require elucidation.

Note each mass has implicit e and π factors due to

$$\nu'_{new} = \nu_{e\ new}\ (\alpha_0{}^2)^{-1} \cong e\pi\ \times 10^{-5}\ \text{GeV}/c^2 \tag{2.16}$$

2.5.2 The Mass Sequences of the Sequences Table

The Sequences Table of Fig. 2.5 has entries for each of these fermion mass sequences. It also has entries for the strata and Gambol fermion mass sequences:

C Fermion Sequence 2 Masses	1	$2/3 \times 2^5 = \ln_2((32^{32})^{0.133})$	1.3	1.3
D Fermion Sequence 1 Masses	$½^{½}$	$3/2 \times \pi 2^5 = \ln_2((32^{32})^{0.3\pi})$	0.65	0.81
E Fermion Strata Masses Sequence	½	$95939 = 5.96\ \ln_2(32^{32\cdot 32\pi})$	0.325	0.325
F Fermion Gsmbol masses factors based on α_{QED}	$0.48 \approx 2^{½}/3$	$844242 = 5.15\ln_2(32^{32\cdot 32\cdot 32})$	0.299	0.303

The entries in these rows show an appealing, and unexpected, regularity in their use of series of factors of 32 within their logarithmic expressions 32, 32·32, 32·32·32. The sequence is based on n_φ = ½, which is the sequence of sequences of row G: namely 4, 2, 1, ½. The Fth row is based on the fine structure constant α_{QED}. All these sequences follow from the Master Equation solutions for various values of n_φ. We therefore suggest these sequences have a reality although much of it remains to be understood. (The spectrum of strata and Gambol Fermions is symbolized by Fig. 2.3.)

2.6 The λ Values for Cosmos Theory are Types of Coupling Constants

An examination of Fig. 2.5 suggests that the values of λ_{n_φ} are effectively coupling constants, Row E shows the value $\lambda_{n_\varphi=½}$ = 0.325 which is the Coupling Constant for the Fine Structure Constant:

E Fermion Strata Masses Sequence	½	$95939 = 5.96\ \ln_2(32^{32\cdot 32\pi})$	0.325	0.325

We now turn to investigate the λ values of the fermion masses expressed in terms of the Coupling Constants in Fig. 2.6.

2.6.1 The λ Value for Cosmos Theory Power Sequence of Spaces

The Cosmos Theory set of spaces has a λ value for the doubling of the length of a dimension array column:[22]

$$d_{cd} = 2^{r/2 + 2} \tag{2.17}$$

For n_φ = 4 we set

$$2 = \exp(\pi/(\lambda_4 - ¼)^{½})$$

[22] The fractal Cosmos Curve construction is based on column lengths. See Blaha (2024b) and (2024c).

Solving for λ_4 we obtain

$$\lambda_{n\varphi=4} = 16/m = \lambda_4 = 20.79 \tag{2.18}$$

for $n_\varphi = 4$. As a result, eq. 2.18 implies

$$m = 16/\lambda_{n\varphi=4} = 0.7696 \tag{2.19}$$

and

$$m \approx \pi/4 = 0.785 \tag{2.20}$$

We will use the value of m = 0.7696. Note m is dimensionless.

2.6.2 The λ Value for the Coupling Constant Power Sequence

The Coupling Constant series has a multiplicative power of four embedded in the known coupling constants. We note

$$4 = \exp(\pi/(\lambda_{n\varphi=2} - ¼)^{½}) \tag{2.21}$$

generates the Cosmos coupling constant spectrum. As a result eq. 2.19 gives

$$\lambda_{n\varphi=2} = [(\ln(4)/\pi]^{-2} + ¼ = 5.386 \tag{2.22}$$

2.6.3 The λ Value for Fermion Mass Power Sequences

The approximate masses for the two fermion sequences appear in Fig. 2.9. Taking the fermion progression factors seriously we can calculate the λ values, which relate to n_φ, for each series:

$$\lambda_{n\varphi seq1} = [(\ln(32\pi)/\pi]^{-2} + ¼ = 0.714 \qquad \approx g_2 = 0.63 \tag{2.23}$$
$$\lambda_{n\varphi seq2} = [(\ln(32)/\pi]^{-2} + ¼ = 1.07 \qquad \approx g_3 = 1.21 \tag{2.24}$$

where the coupling constants, g_2 and g_3, shown for comparison, are taken from Fig. 2.6. Note the similarity of coupling constant values to the λ values. We describe the basis of this similarity later.

If we set $n_{\varphi 2} = 1$ and $n_{\varphi 1} = ½^{½}$ we find

$$\lambda_{n\varphi=seq1} = n_{\varphi 1}^{\;2}/m = 0.65 \tag{2.25}$$
$$\lambda_{n\varphi=seq2} = n_{\varphi 2}^{\;2}/m = 1.3 \tag{2.26}$$

Thus there is an approximate match between fermion mass sequence factors and coupling constants:

$$\lambda_{n\varphi=seq1} = 0.65 \approx g_2 = 0.63 \tag{2.27}$$
$$\lambda_{n\varphi=seq2} = 1.3 \approx g_3 = 1.21 \tag{2.28}$$

These values are approximately correct to 3% for sequence 1 and 7.4% for sequence 2.

The multiplicative factor for sequence 1 is

$$g_{seq1} = \exp(\pi/(\lambda_{n_\varphi=seq1} - {}^1\!/\!_4)^{1/2}) = 143.64 \qquad (2.29)$$
$$= 1.43 \times 32\pi \approx 3/2 \times 32\pi$$

and for sequence 2 is

$$g_{seq2} = \exp(\pi/(\lambda_{n_\varphi=seq2} - {}^1\!/\!_4)^{1/2}) = 21.45 \qquad (2.30)$$
$$= 2/3 \times 32$$

thus matching the fermion mass growth factors fairly well in Fig. 2.9. The factors of 3/2 and 2/3 are suggestive but their understanding must await a deeper analysis.

2.6.4 The Fundamental Fermion Masses Model[23]

The mass sequences, as pictured above, have a geometric character similar to the energy levels and the Cosmos spaces spectrum. We now define a 2×2 scaling equation, similar to the Master Equation, for *both* sequences.

$$\begin{bmatrix} d^2/dr^2 + \lambda_1/r^2 - M_1 & 0 \\ 0 & d^2/dr^2 + \lambda_2/r^2 - M_2 \end{bmatrix} \begin{bmatrix} \varphi_1 \\ \varphi_2 \end{bmatrix} = 0 \qquad (2.31)$$

We related the multiplier of each sequence to a coupling constant as shown earlier:

$$2^5\pi = \exp(\pi/(\lambda_1 - {}^1\!/\!_4)^{1/2}) \qquad (2.32)$$
$$2^5 = \exp(\pi/(\lambda_2 - {}^1\!/\!_4)^{1/2}) \qquad (2.33)$$

for the up-type and down-type sequences respectively. Then

$$[(\ln(32\pi)/\pi]^{-2} + {}^1\!/\!_4 = \lambda_1 = 0.714 \approx g_2 = 0.63 \qquad (2.34)$$

$$[(\ln(32)/\pi]^{-2} + {}^1\!/\!_4 = = \lambda_2 = 1.07 \approx g_3 = 1.21 \qquad (2.35)$$

where we adjoin Coupling Constant g values of comparable value from the coupling constant table (Fig. 2.5).

Note the values of λ_1 and λ_2 are related by:

$$1.07/.714 = 1.5 \qquad (2.36)$$

Note the sum of the values $\lambda_1 + \lambda_2 = 1.784$ and also $g_2 + g_3 = 1.84$. These values differ by 3.4% suggesting that the ElectroWeak SU(2) and Strong SU(3) interactions combine to generate each mass sequence through an interplay of interactions.

---

[23] Blaha (2025b).

2.6.5 Separation of SU(2) and SU(3) Contributions to the Sequences' Powers

We view the values in the sequences as sums of SU(2) and SU(3) terms. We calculate the discrepancy x between the λ_i from

$$\lambda_1 + x = 1.21 \qquad (2.37)$$
$$\lambda_2 - x = 0.63$$

using the approximate equality: $\lambda_1 + \lambda_2 = 1.784$ while $g_2 + g_3 = 1.84$. We find

$$x = 0.468 \qquad (2.38)$$

.After noting that $g_4 = 2.4$ is itself an extrapolation we will use the approximate values from Fig. 2.5

$$x = ½(g_1 + g_2) = 0.466 \qquad (2.39)$$

Then we find an approximate relation between the SU(1), SU(2) and SU(3) coupling constants, and the λ_i, factors in in the fundamental fermion mass sequences.

$$\lambda_1 = 0.714 = 1.21 - 0.468 = g_3 - ½(g_1 + g_2) \qquad (2.40)$$

for sequence 1, and

$$\lambda_2 = 1.07 = 0.63 + 0.468 = 3g_2/2 + ½\, g_1 \qquad (2.41)$$

for sequence 2.

Sequence 2 is governed by the ElectroWeak interactions. Sequence 1 is governed by a combination of ElectroWeak and Strong SU(3) interactions. These results are reflected in the data since sequence 1 masses grow much more rapidly than sequence 2 masses.

The reasonableness of the above expressions for the multipliers of the sequences supports the choice of the sequence 1 multiplier $2^5\pi$ based on the t/c mass ratio, and the sequence 2 multiplier 2^5 approximately based on sequence 2 masses. They approximate the values in Fig. 2.5. The dependence of the multipliers of interactions would appear to be directly useful for specifying the dynamics of gambols in the author's Gambol Model presented in previous books.

It is important to view the generation of fermion masses with Strong and ElectroWeak interactions as occurring *within* each individual fermion. The Higgs mechanism is used conventionally to generate vector boson masses that may play a role within interaction effects in fermions.

However the success in describing the fermion masses raises the question whether vector boson masses may also be similarly described. In addition, the masses of scalar bosons may also be so described. Previously we have shown numerical relations between vector and scalar boson masses in Blaha (2024g) that are suggestive of this possibility.

Thus this section directly relates fermion mass sequences to the coupling constant sequence.

2.7 Fermion Strata

The fermion strata, where each fundamental fermion becomes four identical fermions except for stratum number, were described in earlier books and symbolized in Fig. 2.3. Their sequence is specified in row E of Fig. 2.5.

E Fermion Strata Masses Sequence	½	$95939 = 5.96\ \ln_2(32^{32\cdot 32\pi})$	0.325	0.325

The n_φ value is ½ continuing the sequence of its values: 4, 2, 1, ½. The multiplicative factor for masses is 95939. It is the ratio of the masses of the four almost-identical fermions generated for each initial fermion. At this point in time we only know of the lowest mass stratum for the fundamental fermions due to the large masses of their partners.

The λ value, 0.325, is ½ of the sequence 1 λ value of 0.65 in eq. 2.25. Thus we can represent it as a combination of SU(1), SU(2) and SU(3) coupling constants

$$\lambda_{\text{strata}} = \tfrac{1}{2} g_3 - \tfrac{1}{4}(g_1 + g_2) \qquad (2.42)$$

2.8 Gambol Fermions

The Gambol Fermions, which each correspond to a fundamental fermion, is a partner of the fundamental fermion with the same internal symmetry quantum numbers. Each gambol also has a separate gambol number. The gambol spectrum is symbolized in Fig. 2.3. A gambol has a mass that is a factor of 1/844242 of the mass of its corresponding fermion. The gambol sequence is specified in row F of Fig. 2.5.

F Fermion Gambol masses factors based on α_{QED}	$0.48 \approx 2^{1/2}/3$	$844242 = 5.15\ln_2(32^{32\cdot 32\cdot 32})$	0.299	0.303

The n_φ value is 0.48. It was set by the value of the Fine Structure Coupling: $g_\alpha = 0.303 = (4\pi\alpha)^{1/2}$. At this point in time we have no direct evidence of gambols, which appear to be confined within fundamental fermions. There is much indirect evidence for gambols. See Blaha (2023e) and (2024a). Gambols may be likened to confined quarks – but at a much deeper level.

The λ value, 0.303, is g_α, the U(1) coupling constant, indicating gambol mass is based on QED.

2.9 Experimental Confirmation of Cosmos Theory in Our Universe

We have seen that the structure of Cosmos Theory space for our universe describes the patterns of symmetries and particle masses. (We showed the beginnings of the understanding of boson masses in previous books.) We have showed the known Coupling Constants are accurately represented by a Cosmos Sequence. We have shown the fundamental fermion masses are approximately specified by two Cosmos sequences. We have related mass sequence multipliers to coupling constants.

Therefore we see strong evidence for the validity of Cosmos Theory when applied to our universe and, by extension to other Cosmos universes. In particular Cosmos Theory gives the Unified SuperStandard Theory[24] (UST) which is an extension of The Standard Model, as the theory for our universe.

[24] See Blaha (2018a) and (2022f) for the UST.

3. Sequential Derivation of Cosmos Theory

The structure of Cosmos Theory consists of four interrelated parts. It can be viewed as being based on the exact Master Equation or the Extended Master Equation. The Master Equation is independent of interactions and thus independent of quantum corrections. (The $1/r^2$ potential is a self-potential – not an interaction.)

The structure of the four Cosmos Theory components is "classical" with no interactions and no quantum corrections. Our previous books have considered the following in detail: The Cosmos Fractal Curve, The Cosmos Spaces Spectrum, The Extended Master Equation, Sequence Theory, Cosmos Space Fractals and HyperSpheres, Mass-Energy and Child Universes, and the origin of energy in motion.

3.1 Quantum Cosmos

The Parent universe is quantum within the 3D Cosmos Space. Its wave function is developed in chapter 4 of Blaha (2025j) Chapter 5 discusses implications of its Uncertainty Condition:

$$[E, t] = i\hbar \qquad (3.1)$$

where for Cosmos Space we choose $t = 2^n$ = the length of the piece-wise linear segment where n is the Cosmos Fractal Space construction step. The energy[25] $E = d_{dr}$ = dimension array size = number of dimension array fermions of the Parent universe that appear in its quantum field in eq. 4.22 below:

$$\psi_{n_\varphi}(\varphi, u, t) = \sum_n \int dk\, N(k, n)\{a_{n_\varphi}(k, n)\, e^{-iEt - ik\varphi - i\lambda X - i\lambda P} + a_{n_\varphi}^\dagger(k, n)\, e^{iEt + ik\varphi + i\lambda X + i\lambda P}\} \qquad (4.22)$$

from Blaha (2025j).

The conjugate variable of E is t. We determined t from the running length of the Cosmos Fractal Curve measured in construction steps n. Thus we may[26] take $t = 2^n$. We have also identified each construction step n within a Cosmos space with a dimension array size of 2^{r+4} where r is the space-time dimension and $n = r + 4$.[27] As a result

$$[E, 2^n] = [d_{dr}, 2^n] = i\hbar \qquad (3.2)$$

in the 3 dimension Cosmos Space. The forms of eq. 4.22 above and eq.3.2 suggest an alternate form:

$$[d_{dr}, n] = [d_{dr}, r] = i\hbar \qquad (3.3)$$

---

[25] This energy is not the E appearing in the evaluation of the Master Equation solutions.

[26] The line length in step n is 2^n. We will generalize n to be continuous in this section.

[27] For 2nd Kind HyperCosmos spaces we set $n = r + 3$ and set the array size to 2^{r+3}. Introducing fermion strata and gambol fermions would change n to r+7 and r+6 for HyperCosmos and 2nd Kind HyperCosmos spaces respectively.

where n = r + 4 (HyperCosmos spaces) or r + 3 (2nd Kind HyperCosmos spaces), and where the energy and construction step number n and also dimension r, made into continuous variables, satisfy an uncertainty relation. Then a measurement of the energy (number of dimension array fundamental fermions) makes the construction step variable and the dimension uncertain; and *vice versa*. Thus we are on the threshold of *quantum* fractal curves. See chapter 7.

Dimension array size and dimension r satisfy an uncertainty condition.

If one is determined then the other becomes uncertain for a quantum fractal determination of Cosmos spaces. One can compare this situation with the coordinate – momentum uncertainty condition: $[p, x] = i\hbar$. Similarly d_{dr} can be represented by

$$d_{dr} \equiv i\hbar \, d/dr \tag{3.4}$$

The door is thus opened to a *Quantum Cosmos Theory*, which bears further investigation. In this theory spaces, and universes, are quantized. An initial effort has been made with the quantization of the Parent universe. Its quantum field is a solution of the (Extended) Master Equation.

Chapter 7 considers Quantum Cosmos Theory.

3.1.1 Quantum Child Universes

Child universes are quantum within the Parent universe and within their interiors. The Parent universe and child universes have space-time coordinates within themselves and thus experience renormalization due to inherent interactions within them.

3.2 Coupling Constants

The Coupling Constant sequence (Fig. 2.6) that we have found is determined by the Master Equation. This equation is quantum within the Cosmos Space. However it has no external interactions or fields. Thus there is no renormalization of the Coupling Constant sequence values. \

When they are matched to experimentally determined coupling constants, which are known at "low" energies, one finds the match is exact to within experimental error. Interestingly, the experimentally determined values contain corrections that may be calculated in Perturbation Theory. These corrections introduce a momentum dependence in the coupling constants.

The Coupling Constant Sequence values are accurate as far as we know at known energies.

3.3 Fermion Masses

The masses of the fundamental fermions were found to be approximately determined in two sequences. (Fig. 2.9) There were some significant deviations in the sequences. We attribute these deviations, which occur with our child universe, to perturbation theory effects that we considered provisionally in chapter 2.

3.4 Higher Dimension Spaces and Two Tier Quantization

Perturbation Theory has been plagued with divergences for many years. Renormalization techniques have attempted to eliminate the divergences Physically. However these techniques are limited to our four dimension space time and cannot be applied to the higher dimension universes that are part of Cosmos Theory.

In 2002 this author developed a Quantum theory called Two Tier Theory[28] that *removed any divergences in a space-time of any number of dimensions for all polynomial Lagrangian interactions.* It was based on a simple mechanical model of a harmonic oscillator residing on another harmonic oscillator. Thus the two tiers of oscillators.

\Two Tier Theory, when applied in our universe, gives *identical* results, to great precision, to the usual Quantum Field Theory Perturbation Theory for known energies. At asymptotically high energies its results are finite. (This observation applies to the famous three fermion triangle diagram.) Thus Two Tier Theory gives:

1. Elimination of Perturbation Theory divergences in any number of space-time dimensions for any polynomial Lagrangian model.

2. Effectively exact agreement with known perturbation theory results.

Cosmos Theory universes have no divergences in any universes if Two Tier Theory is used. See Blaha (2002) and (2005a).

3.5 No Renormalization of Cosmos Structure

The structure (spaces) of Cosmos Theory is determined by the Master Equation and the other parts of Cosmos Theory. The Cosmos spaces and their dimension arrays are specified *before* any space-times and before any interactions appear. (Interactions exist only within the Parent universe and child universes since they are "derived" from the dimension arrays of these entities.) *There is no renormalization of Cosmos structure.* These points were made in our previous books.

---

[28] Blaha (2002) and (2005a).

4. Beyond the Primordial Point

The primordial point of Cosmos Theory may be viewed as a dimensionless Mathematical point of zero extension – no breadth, no width, and no lifetime. It does not exist in a space (or space-time). It cannot be said to have a beginning or an end. It coexists with all the other components of the Cosmos although we describe it as the source of the other components: the Cosmos Fractal Curve, the Cosmos space of three dimensions, and the Extended Master Equation residing in the Cosmos Space.

The primordial point raises a number of questions: Is there only *one* primordial point? Answer: Yes. Where is it located? Answer: undefined since there is no space and thus no coordinates. (The Cosmos Space arises from the Cosmos Fractal Curve – the unit square it generates that scales to the CosmosSpace spatial coordinates and the Cosmos Space's time, which is the running length of the Cosmos Fractal Curve as it is being constructed. What is the primordial point's lifetime? Answer: It has no Beginning and no end. Thus immortal.

The above description raises the question of the nature of the primordial point. One may say it is a Mathematical point. But is there more? In view of its lack of features one can associate a further feature with the primordial point – we may choose to identify it as a sole non-zero dimension (without numerical value) in the 180 dimension Penultimate Unification Space described in earlier books. In Blaha (2025j) we showed how such a dimension can evolve (in accord with Cosmos Entropy) to the full set of Cosmos spaces. This view introduces the possibility of a self-consistent, self-generation. The results are the same as in the sequential generation of the Cosmos from a single Mathematical point.

It is also possible to view the primordial point as the result of an "earlier" stage of "creation" with an alternate Beginning.

4.1 Possibility of a Primordial Point Substructure

We have seen that Cosmos Theory may be derived from a single primordial point. It is possible to consider the primordial point is the consequence of a prior Physics. (We have chosen to view the primordial point as acted upon by an Agent (God) for the generation of the Cosmos.)

If we play the Devil's advocate and consider the possibility of a prior origin of the primordial point, then certain considerations become relevant. The prior Physical situation must apparently be a space with perhaps an active dynamics and vacuum. It is possible the primordial point surfaces due to a chaotic evolution. In any case a dynamics presupposes a coordinate system for the prior space. In particular, it presupposes a time evolution before the appearance of the primordial point. Now the question of the nature of the primordial point surfaces: Is there only one such point? If there are more than one point, then there are many questions: Are there many

Cosmoses? How did our Cosmos's mathematical point appear? Is it really a Mathematical point? How do the spatial and time coordinates of the prior space surface in our Cosmos (since there is no need for them in our Cosmos Theory)? And so on.

Perhaps the most telling negative of positing a dynamics prior to the primordial point is raised by the question? Where did that prior dynamics arise? It appears that going prior to the primordial point opens the possibility of a chain of dynamics that is not needed – it does not extend Cosmos Theory – it leads to a series of pasts of the Cosmos without justification. *In the end it also would require an Agent as the ultimate cause.*

Therefore as an expedient principle of Economy we choose Cosmos Theory as we have developed it with a single primordial point. It is sufficient to obtain Cosmos Theory, and Physics as we know it. The alternative of an "eternal recurrence" of previous Cosmic cycles is unnecessary and serves no Physical purpose.

5. Cosmos Theory View of God

We have considered the question of an Agent that initiates the Cosmos and "manages" the evolution of universes. We now view the Agent as God and discuss God's Physical relation to the Cosmos. In Blaha (2025j) we found the only reasonable way to initiate the Cosmos was through the Agent. Any other choice would lead to an infinite chain of previous origins. Since the single mathematical point that initiates the Cosmos leads to all reasonable known Cosmos features, there is no need for a precursor to the Cosmos point. We therefore found that the Agent, which we can only view as God, is the sufficient initiator of the Cosmos. We now consider a Physical view of God.

5.1 Physical Concept of God

There are many views of God. In this work based on Cosmos Theory we will view God from a Physical perspective. What does the Physics of Cosmos Theory tell us about the Physical aspects of God? (We will not consider the other aspects of God such as His mercy, kindness, impact on Humanity, and so on that occupies theological discussions of Divinity.)

We view God in terms of His impact on the Cosmos within the framework of Cosmos Theory. First, we view God as the Agent of the "start" of the Cosmos from a single point realizing the Cosmos has no beginning or end, and thus coexists with God. In considering this role we differentiate God from the Cosmos viewing the Cosmos as the Creation.

Secondly, we view God as the Agent of changes in the universes of the Cosmos. Unlike the first role, which is "static" in nature, the second role has God as dynamically active in effecting changes in universes – including our own universe. Changes can take two forms from a Physics viewpoint: one form is to conform to the Laws of Physics, which He instituted, and only "tweak" quantum processes to effect changes. The "tweaks" fall within the quantum probabilities of Physical evolution and thus do not violate Physical process evolution. They occur "silently" but can cascade with macroscopic consequences. For example, a tweak on an electron in a water vapor cloud can cause a condensation that precipitates a bolt of lightning that fells a tree and possibly kills a person or cow and so on.

Another form changes may take is macroscopic violations of Physical Laws. We call events of this form "*miracles*." Given the variety of Physical phenomena, miracles may happen "here or there" without long term Physical implications but with possibly long term human related (spiritual) effects.

Thus Cosmos Theory accommodates a cause of the Cosmos as an action of the Deity and the management of universes as an attribute of God.

5.2 God as the Ultimate Quantum Observer

In our previous books we broached the question of a Quantum Cosmos Theory that begins at the stage of the Cosmos Fractal Curve and the Cosmos spectrum of spaces. Quantum Cosmos Theory has been discussed in some detail in Blaha (2025i).

Quantum Theory assumes the presence of an observer for the analysis of Quantum phenomena. The quantum formulation of Cosmos Theory thus requires an observer for all components.

There is only one choice for a Quantum Observer at all stages of the Cosmos evolution: God. Thus we specify God is the *Ultimate Quantum Observer*. This choice carries over to Quantum phenomena in universes such as ours. Therefore apparent paradoxes involving Quantum observers such as the Schrödinger Cat Paradox are resolved when one includes the ever-present silent observer – God. Thought experiments, such as questioning whether events occur in the absence of an observer, are resolved by the implicit presence of God, the Observer.

We conclude *God "makes" Reality*. All Natural events are observed. As a consequence the chart of the components of Cosmos Theory should add the ultimate Observer God, the initiator of the Cosmos, as the Fifth component that was present at the initiation of the Cosmos. The introduction of God into Quantum Cosmos Theory reflects the Simplicity and Serenity of its formulation as Reality. Would God have done otherwise?

5.3 Features of God in Relation to Cosmos Theory

We now turn to consider the relation between God's features and corresponding Cosmos Theory features. We inquire of the extent that Cosmos Theory "models" God.

Qualities of God *not* apparent in Cosmos Theory

These qualities of God do not have a Physical analogue in Cosmos Theory.

Intelligence
Existence as a Being
Emotions and Personality
Self-Knowledge
Freedom of action
Spirituality
Senses of the surroundings
Communicability

Qualities of God apparent in Cosmos Theory

Omnipresence
No beginning or end
Omniscience

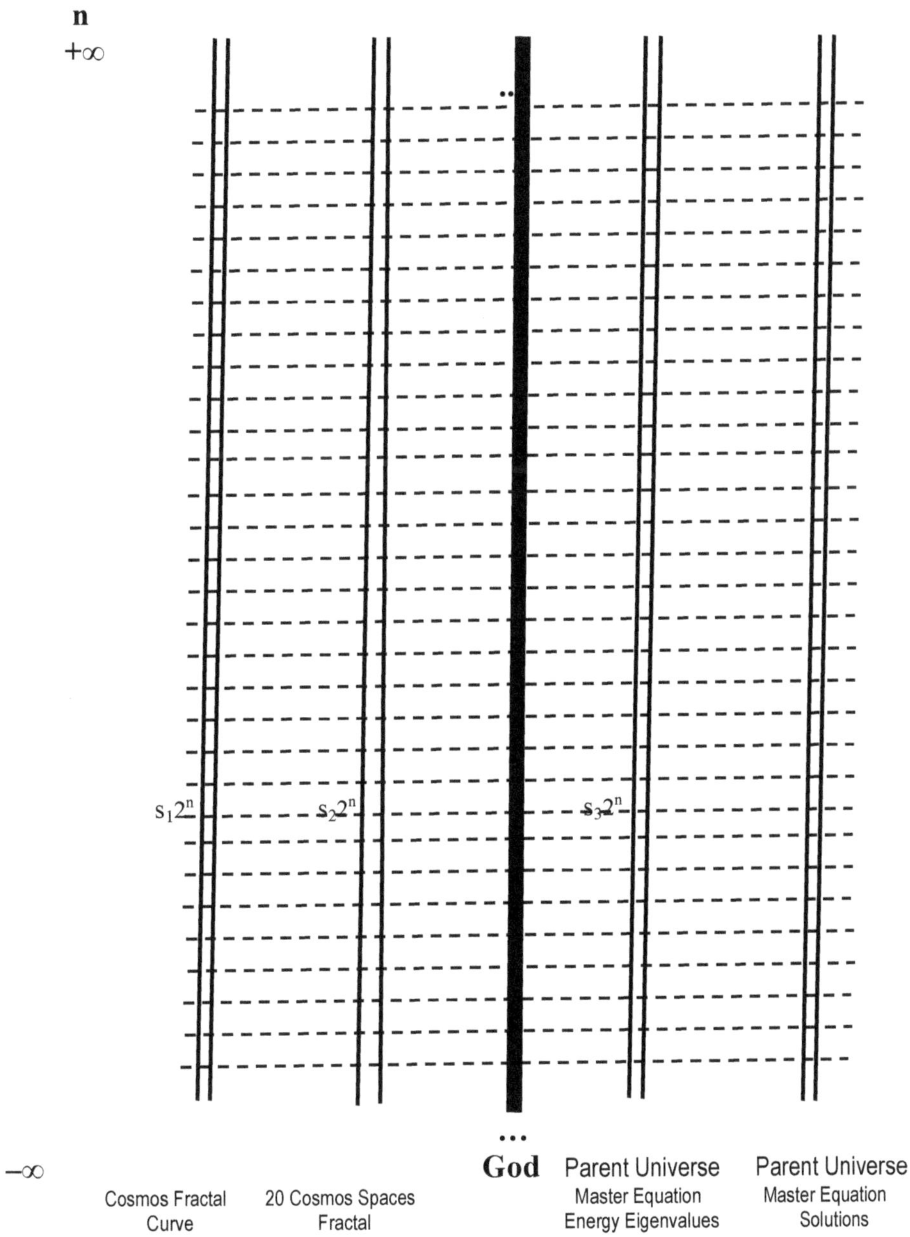

Figure 5.1. The *five* components of the Cosmos with the addition of God as the initiator of the Cosmos and as coexistent with the Cosmos.

6. No End to the Cosmos

Having considered the Physical nature and role of God as the initiator of the Cosmos, we now consider a broad view of His creation.

6.1 The Cosmos as a Concept

Some 6 years ago, having developed the Unified SuperStandard Theory (UST), we began the study of the Cosmos using Cayley numbers after realizing that the numerics of the UST clearly indicated their relevance. We developed Cosmos Theory as a result.

We now ask: Where has Cosmos Theory led us? We have seen the remarkable match of Cosmos Theory with known experimental results for Elementary Particles: the structure of interactions, their coupling constants, the structure of fundamental fermions, and fundamental fermion masses. The only remaining, but important, issues are the calculation of particle scattering amplitudes and some constants such as magnetic moments, These quantities can be determined based the above Cosmos results although their calculation is challenging. The fact that they are determined is the key result.

The nature of Cosmos Theory as an extension of The Standard Model is particularly interesting. The Standard Model and the UST are based on "nitty-gritty concrete" data about particles and interactions. Cosmos Theory brings in new conceptual considerations. It begins with a point with no properties. This point might be considered a dimension – a primitive entity. Then it derives the structure of the spaces of Cosmos Theory – a concept again. With the derivation of the Parent universe and the introduction of child universes concrete, material, entities are formed.

It is clear that the drift of this theoretical development is a trend from the concrete to the conceptual. This trend is encouraging if one considers it in the light of the initiator, God. It appears more reasonable that growth from a concept is more acceptable than growth from a "clump" of particles with mass-energy in view of the spiritual nature of the initiator, God. *One might say that Cosmos Theory is a thought in the mind of the Initiator,*

The essence of the growth of Cosmos Theory lies in the generation of dimensions, to which we now turn,

6.2 Dimensions are the Key

Dimensions are simple things. Each dimension is basically a "place holder" for various quantities such as internal symmetry representation dimensions or fermions. In themselves dimensions are featureless.

Cosmos Theory takes a primitive point or dimension and generates dimensions in sequences. The process is entropic as we showed in detail in Blaha (2025j). Are any material things constructed in this procedure? No, not concrete entities. But the increase in the sets of dimensions leads to the possibilities of structuring (maps) that form concrete items.

The initial stages of Cosmos theory generate abstractions. Before the introduction of universes we find only pure thought unimpeded by the material.

The reliance on the generation of dimensions that eventually are used in the creation of material entities – energies and masses – causes Cosmos Theory and the Cosmos to originate conceptually. Subsequently material entities are created leading to the concrete manifestations that we call the "material" universes.

We thus see an Initiator – God – a pure spirit – start with conceptual entities – thoughts – and then progress to the creation of the material universes. Cosmos Theory is the bridge between the spiritual and the material, which we call Reality.

Cosmos Theory bridges the great divide between Concept and Reality.

6.3 Time, God and the Cosmos

We developed Cosmos Theory with descriptions of stages starting from a point to universes. In fact, all Cosmos components up to the Parent universe exist instantaneously eternally. It sounds paradoxical but the mathematics of Cosmos Theory embodies the instantaneity of the execution of all stages.

God exists instantaneously, eternally, as well. He is beyond time since time does not exist for Him outside of universes. Time does not exist for anything outside universes.

One might ask what God sees within universes. There is only one valid answer: He sees the complete, totally detailed, history of each universe from beginning to end in a manner similar to a complete viewing of a tape. He is not limited by past, present or future.

This observation leads to the issue of Predestination. If God sees all from the beginning to the end of a universe, does that imply Predestination, and the elimination of Free Will? The answer is simple: The progress of a universe embodies Free Will decisions that God sees in the "tape." Knowledge of Free Will decisions does not constitute a restrictive Predestination. *God's complete knowledge of the events of a universe's history is "merely" observational – not compulsory.*

6.4 A "Drift" of Cosmos Theory – towards a Simulation?

The consideration of the relation of time to the Cosmos components and to God has led us to consider the nature of the connection between Cosmos Theory and Reality.

We have seen an increasing tendency to increasing abstractness in the Cosmos and in Cosmos Theory. Cosmos Theory is a concept that is reduced to printed form in our books. It can be viewed as merely descriptive and providing logical connections between the various components of the theory.

One can also view Cosmos Theory as describing "Reality" although it is clearly a representation of Thought and not material. One can go further and suggest Cosmos Theory is a simulation of Reality. We define a simulation as a representation of a Real process or series of events. Cosmos Theory, with that definition, is a simulation. The derivation from a mathematical point is a sequence of events that embodies Reality (modulo its instantaneous nature) – thus a simulation of the origin of the Cosmos although it is not the Cosmos.

We see Cosmos Theory as an example of pure thought that straddles the great divide between Theory and Reality. The interconnections of the components of Cosmos Theory could be called "the widening gyre of Reality."

The Cosmos as expressed by Cosmos Theory can be viewed as the exercise of Spirit that leads to the Material.

7. Quantum Cosmos Theory

In section 5.2 we saw God must be the Ultimate Quantum Observer. With this view, we can take the Cosmos Fractal Curve and extend it to be Quantum: the *Cosmos Quantum Fractal Curve*. This curve must be generated by a dynamic equation which has a form analogous to the Master Equation. It will define a set of Cosmos Quantum Fractal Curve states which can be used to specify the Cosmos spaces.

The role of God in this view is to act as the Quantum Observer that selects the Cosmos state. "God separated the waters creating the firmament" – the Cosmos spaces. *For Cosmos Theory to be truly Quantum there must be a Divine Observer.*

The Quantum selection process is analogous to the role of an observer in the Schrödinger Quantum Cat model who examines the state of the cat thus deciding its status as alive or dead.

7.1 Cosmos Quantum Fractal Curve Master Equation

A Master Equation exists for the Cosmos Quantum Fractal Curve that differs in details from the (Extended) Master Equation for the Parent universe quantum field described in Blaha (2025h) and (2025i):

> It is a Schrödinger-like equation with a $1/r^2$ potential. The equation scales and allows us to avoid the introduction of a mass scale. The scaling $1/r^2$ potential has the advantages:
>
> 1. No mass scale dependence.
> 2. A spectrum of powers of a multiplier times an overall constant.
> 3. A constant that can be scaled out of the Master Equation by scaling the radial coordinate.
> 4. The capability to scale any energy eigenvalue to one.
> 5. The generality of the scaling $1/r^2$ Master Equation solution is supported by its solution also being equivalent to a harmonic oscillator-like solution and also a Free Field solution for small values of the radial parameter.
>
> Perhaps the most significant feature of the $1/r^2$ potential is that it generates a set of eigenvalues in the form of a sequence of powers of a constant. This feature corresponds to the power series of the Cosmos spaces dimension array sizes.

A scaling potential provides the simplest guarantee of an energy spectrum that is a sequence of powers of a constant.

We now note the Master Equation for the Cosmos Quantum Fractal Curve need only have one "spatial: coordinate x.[29] It also has one time coordinate – the running length of the fractal curve. Then this Master Equation is[30]

---

[29] There is only one spatial coordinate since the Cosmos Quantum Fractal Curve creates a one dimension unit line. In contrast a classical fractal curve creates a two dimension unit square. The Cosmos Quantum Fractal Curve is an infinite line of Hausdorff dimension one.

$$[\partial^2/\partial x^2 + \lambda/x^2 \ - \partial^2/\partial t^2]\psi(x, t) = 0 \qquad (7.1)$$

which becomes

$$[\partial^2/\partial x^2 + \lambda/x^2 + C)]\psi(x) = 0 \qquad (7.2)$$

where

$$\psi(x, t) = \psi(x)e^{-iEt} \qquad (7.3)$$

and $C = E^2$. Solving eq. 7.2 we find

$$C_n \ = c \exp(-2\pi n/(\lambda \ - 1/4)^{1/2}) \qquad (7.4)$$

and

$$E_n = C_n^{\ 1/2} = c^{1/2}\exp(-n\pi/(\lambda \ - 1/4)^{1/2}) \qquad (7.5)$$

where n can be a positive or negative[31] integer. Negative values of n correspond to fractal construction steps. Positive values of n correspond to the steps that create a unit line from a point. Thus our Cosmos spaces' generation process takes a point to an infinite one dimension space. Section 2.2 of Blaha (2025f) describes the construction process:

> The Cosmos Curve begins with a zero dimension point, and grows to a one dimension piecewise linear line segment by a simple aggregation of infinitesimal elements based on the identity
>
> $$\sum_{k=\infty}^{1} 2^{-k} = 1 \qquad (2.17)$$
>
> It starts from a point where $k = \infty$, initially adding infinitesimal increments, which then sum to a unit line segment. The order of the summation begins with the infinitesimal and proceeds to the larger summands.
>
> Then the unit piecewise linear line segment grows through a construction consisting of steps to form a fractal line. At the nth stage of the construction the growing line has the length 2^n. … The line becomes fractal in the limit $n \to \infty$.

The value C_n is the size of the piecewise linear line segment of the nth construction step. By eq. 7.4 it can be expressed as

$$C_n = 2^n \qquad (7.6)$$

when we let $n \to -n$ (for convenience) and we set

$$\exp(2\pi/(\lambda \ - 1/4)^{1/2}) = 2 \qquad (7.7)$$

to obtain a Hilbert-like fractal curve. Eq. 7.7 implies

$$\lambda = \ (2\pi /\ln(2))^2 + 1/4 = 82.42 \qquad (7.8)$$

---

[30] Note the resemblance to a massless Klein-Gordan equation with a self-interaction.

[31] The eigenvalue spectrum results from the imposition of orthogonality condition on the solution function near $x = 0$ according to von Neumann. The value of C_n can be taken to be positive since its sign can be changed by Wick rotations of the time variable: $t \to it$.

for the Cosmos Quantum Fractal Curve.

7.1.1 Cosmos Quantum Fractal Curve Quantum Field

A solution of eq. 7.2 has the form

$$\psi_{E,\,n}(u, t) = e^{iEt} \cos((\lambda - ¼)^{½}\ln(2^n x)) \qquad (7.9)$$
$$= e^{iEt} \cos(\ln(2^n x)^{\lambda})$$

We define the corresponding Quantum field:

$$\psi(x, t) = \sum_n N(n)\{a(n)\exp(-i2^n t)(2^n x)^{-i\lambda} + a^{\dagger}(n) \exp(i2^n t)(2^n x)^{i\lambda}\} \qquad (7.10)$$

We then define

$$p = 2^n \qquad (7.11)$$
$$E = p$$

treating 2^n as a momentum p, and E as the energy. One can view the field as representing a massless boson with self-interaction. The relation of p and n gives

$$dn/dp = 1/p \qquad (7.12)$$

Further defining

$$X = \ln x$$
$$P = \ln p$$

and using

$$\cos(\ln(2^n X)^{\lambda}) = (\exp(i\ln(2^n X)^{\lambda}) + \exp(-i\ln(2^n X)^{\lambda}))/2 \qquad (7.13)$$
$$= ((2^n X)^{i\lambda} + (2^n X)^{-i\lambda})/2$$

we find eq. 7.10 implies[32]

$$\psi(x, t) = \int d(\ln p)\, N(p)\{a(p)\exp(-iEt)(px)^{-i\lambda} + a^{\dagger}(p) \exp(iEt)(px)^{i\lambda}\} \qquad (7.14)$$

or

$$\psi(X, t) = \int dP\, N(P)\{a(P)e^{-iEt - i\lambda X - i\lambda P} + a^{\dagger}(P)e^{iEt + i\lambda X + i\lambda P}\} \qquad (7.15)$$

Note that eq. 7.15 is the Quantum field for the entire Cosmos Quantum Fractal Curve and the raising and lowering operators a and $a^{\dagger}$ create/destroy the Pth construction step n of the fractal curve where P = n.

We now define commutation relation

$$[a(P), a^{\dagger}(P')] = \delta(P - P') \qquad (7.16)$$

We set

[3232] See chapter 5 of Blaha (2025i) for a similar derivation for the Master Equation solution.

$$[\psi(X, t), \psi'(X', t)] = i\,\delta(X - X') \tag{7.17}$$

where

$$\psi_{'}(X', t) = \partial\psi(X', t)/\,\partial t \tag{7.18}$$

to determine the normalization factor N(P). We see the momentum P may be interpreted as indicating a fractal.

We also see a possible commutation relation between a Cosmos space coordinate X and dimension P = n:

$$[X, P] = i\hbar \tag{7.19}$$

This Uncertainty condition asserts the uncertainty of the piecewise linear length in construction step n, and the construction step number n. This condition also applies to a fractal Hilbert curve (which would be a linear curve of Hausdorff dimension one.)

Cosmos Quantum Fractal Curve, the Cosmos spaces, the construction steps, and the Master Equation parameters are all "uncertain". See section 3.1 for a similar uncertainty condition.

7.2 Cosmos Quantum Fractal Curve States

We can define a Cosmos Quantum Fractal Curve step n state with

$$|P> = a^{\dagger}(P')|0> \tag{7.20}$$

and define a total Cosmos Quantum Fractal Curve state with

$$|\text{ Fractal Curve} > \; = \; \int_{-\infty}^{\infty} dP \,|\, P> \tag{7.21}$$

We can consider superpositions of these eq. 7.20 states in a manner similar to that of chapter 5 of Blaha (2025i).

Fig. 1.1 shows the Cosmos Fractal Curve in its double helix form. It combines the HyperCosmos, and 2nd Kind HyperCosmos spaces. It can be made Quantum directly based on the above discussions. Then one has an even more similar form of the Double Helix of Biochemistry.

7.3 The Quantum Cosmos

We now have a complete Quantization of Cosmos Theory:

Cosmos Quantum Fractal Curve.
The (Extended) Master equation and solutions.
The set of Cosmos spaces.
The Quantum Parent universe.
The set of Quantum Child universes.

The structure of the Cosmos consists of the spaces and their associated features, the Master Equations and their solutions, the Cosmos Space, the Parent universe, and the Cosmos Quantum Fractal Curve. These components are independent of any time and thus exist "forever" without beginning or end. Simultaneous! Everlasting! They are all Quantum constructs.

The child universes have creation times, and possibly end times, as measured by the time variable(s) within the Parent universe. Within each child universe a time(s) exists for phenomena within the universe.

Yet all is based on an *ultimate* Quantum Observer, God,
with lesser observers within universes.

REFERENCES

Achiever, N. I., Frank, A. H. (tr), 1962, *The Calculus of Variations* (Blaisdell Publishing, New York, 1962).
Bjorken, J. D., Drell, S. D., 1964, *Relativistic Quantum Mechanics* (McGraw-Hill, New York, 1965).
Bjorken, J. D., Drell, S. D., 1965, *Relativistic Quantum Fields* (McGraw-Hill, New York, 1965).
Blaha, S., 1995, *C++ for Professional Programming* (International Thomson Publishing, Boston, 1995).
_______, 1998, *Cosmos and Consciousness* (Pingree-Hill Publishing, Auburn, NH, 1998 and 2002).
_______, 2002, *A Finite Unified Quantum Field Theory of the Elementary Particle Standard Model and Quantum Gravity Based on New Quantum Dimensions™ & a New Paradigm in the Calculus of Variations* (Pingree-Hill Publishing, Auburn, NH, 2002).
_______, 2004, *Quantum Big Bang Cosmology: Complex Space-time General Relativity, Quantum Coordinates™ Dodecahedral Universe, Inflation, and New Spin 0, ½, 1 & 2 Tachyons & Imagyons* (Pingree-Hill Publishing, Auburn, NH, 2004).
______, *2005a, Quantum Theory of the Third Kind: A New Type of Divergence-free Quantum Field Theory Supporting a Unified Standard Model of Elementary Particles and Quantum Gravity based on a New Method in the Calculus of Variations* (Pingree-Hill Publishing, Auburn, NH, 2005).
______, 2005b, *The Metatheory of Physics Theories, and the Theory of Everything as a Quantum Computer Language* (Pingree-Hill Publishing, Auburn, NH, 2005).
______, 2005c, *The Equivalence of Elementary Particle Theories and Computer Languages: Quantum Computers, Turing Machines, Standard Model, Superstring Theory, and a Proof that Gödel's Theorem Implies Nature Must Be Quantum* (Pingree-Hill Publishing, Auburn, NH, 2005).
______, 2006a, *The Foundation of the Forces of Nature* (Pingree-Hill Publishing, Auburn, NH, 2006).
______, 2006b, *A Derivation of ElectroWeak Theory based on an Extension of Special Relativity; Black Hole Tachyons; & Tachyons of Any Spin.* (Pingree-Hill Publishing, Auburn, NH, 2006).
_______, 2007a, *Physics Beyond the Light Barrier: The Source of Parity Violation, Tachyons, and A Derivation of Standard Model Features* (Pingree-Hill Publishing, Auburn, NH, 2007).
______, 2007b, *The Origin of the Standard Model: The Genesis of Four Quark and Lepton Species, Parity Violation, the ElectroWeak Sector, Color SU(3), Three Visible Generations of Fermions, and One Generation of Dark Matter with Dark Energy* (Pingree-Hill Publishing, Auburn, NH, 2007).
______, *2008a, A Direct Derivation of the Form of the Standard Model From GL(16) (Pingree-Hill Publishing, Auburn, NH, 2008).*
______, 2008b, *A Complete Derivation of the Form of the Standard Model With a New Method to Generate Particle Masses Second Edition* (Pingree-Hill Publishing, Auburn, NH, 2008)
______, 2009, *The Algebra of Thought & Reality: The Mathematical Basis for Plato's Theory of Ideas, and Reality Extended to Include A Priori Observers and Space-Time Second Edition* (Pingree-Hill Publishing, Auburn, NH, 2009).
______, 2009a, *Bright Stars, Bright Universe* (Pingree-Hill Publishing, Auburn, NH, 2008)
______, 2010a, *Operator Metaphysics: A New Metaphysics Based on a New Operator Logic and a New Quantum Operator Logic that Lead to a Mathematical Basis for Plato's Theory of Ideas and Reality* (Pingree-Hill Publishing, Auburn, NH, 2010).
______, 2010b, *The Standard Model's Form Derived from Operator Logic, Superluminal Transformations and GL(16)* (Pingree-Hill Publishing, Auburn, NH, 2010).
______, 2010c, *SuperCivilizations: Civilizations as Superorganisms* (McMann-Fisher Publishing, Auburn, NH, 2010).
______, 2011a, *21st Century Natural Philosophy Of Ultimate Physical Reality* (McMann-Fisher Publishing, Auburn, NH, 2011).
______, 2011b, *All the Universe! Faster Than Light Tachyon Quark Starships & Particle Accelerators with the LHC as a Prototype Starship Drive Scientific Edition* (Pingree-Hill Publishing, Auburn, NH, 2011).
______, 2011c, *From Asynchronous Logic to The Standard Model to Superflight to the Stars* (Blaha Research, Auburn, NH, 2011).

______, 2012a, *From Asynchronous Logic to The Standard Model to Superflight to the Stars volume 2: Superluminal CP and CPT, U(4) Complex General Relativity and The Standard Model, Complex Vierbein General Relativity, Kinetic Theory, Thermodynamics* (Blaha Research, Auburn, NH, 2012).
______, 2012b, *Standard Model Symmetries, And Four And Sixteen Dimension Complex Relativity; The Origin Of Higgs Mass Terms* (Blaha Reasearch, Auburn, NH, 2012).
______, 2013a, *Multi-Stage Space Guns, Micro-Pulse Nuclear Rockets, and Faster-Than-Light Quark-Gluon Ion Drive Starships* (Blaha Research, Auburn, NH, 2013).
______, 2013b, *The Bridge to Dark Matter; A New Sibling Universe; Dark Energy; Inflatons; Quantum Big Bang; Superluminal Physics; An Extended Standard Model Based on Geometry* (Blaha Reasearch, Auburn, NH, 2013).
______, 2014a, *Universes and Megaverses: From a New Standard Model to a Physical Megaverse; The Big Bang; Our Sibling Universe's Wormhole; Origin of the Cosmological Constant, Spatial Asymmetry of the Universe, and its Web of Galaxies; A Baryonic Field between Universes and Particles; Megaverse Extended Wheeler-DeWitt Equation* (Blaha Reasearch, Auburn, NH, 2014).
______, 2014b, *All the Megaverse! Starships Exploring the Endless Universes of the Cosmos Using the Baryonic Force* (Blaha Research, Auburn, NH, 2014).
______, 2014c, *All the Megaverse! II Between Megaverse Universes: Quantum Entanglement Explained by the Megaverse Coherent Baryonic Radiation Devices – PHASERs Neutron Star Megaverse Slingshot Dynamics Spiritual and UFO Events, and the Megaverse Microscopic Entry into the Megaverse* (Blaha Research, Auburn, NH, 2014).
______, 2015a, *PHYSICS IS LOGIC PAINTED ON THE VOID: Origin of Bare Masses and The Standard Model in Logic, U(4) Origin of the Generations, Normal and Dark Baryonic Forces, Dark Matter, Dark Energy, The Big Bang, Complex General Relativity, A Megaverse of Universe Particles* (Blaha Research, Auburn, NH, 2015).
______, 2015b, *PHYSICS IS LOGIC Part II: The Theory of Everything, The Megaverse Theory of Everything, U(4)⊗U(4) Grand Unified Theory (GUT), Inertial Mass = Gravitational Mass, Unified Extended Standard Model and a New Complex General Relativity with Higgs Particles, Generation Group Higgs Particles* (Blaha Research, Auburn, NH, 2015).
____, 2015c, *The Origin of Higgs ("God") Particles and the Higgs Mechanism: Physics is Logic III, Beyond Higgs – A Revamped Theory With a Local Arrow of Time, The Theory of Everything Enhanced, Why Inertial Frames are Special, Universes of the Mind* (Blaha Research, Auburn, NH, 2015).
______, 2015d, *The Origin of the Eight Coupling Constants of The Theory of Everything: U(8) Grand Unified Theory of Everything (GUTE), S^8 Coupling Constant Symmetry, Space-Time Dependent Coupling Constants, Big Bang Vacuum Coupling Constants, Physics is Logic IV* (Blaha Research, Auburn, NH, 2015).
______, 2016a, *New Types of Dark Matter, Big Bang Equipartition, and A New U(4) Symmetry in the Theory of Everything: Equipartition Principle for Fermions, Matter is 83.33% Dark, Penetrating the Veil of the Big Bang, Explicit QFT Quark Confinement and Charmonium, Physics is Logic V* (Blaha Research, Auburn, NH, 2016).
______, 2016b, *The Periodic Table of the 192 Quarks and Leptons in The Theory of Everything: The U(4) Layer Group, Physics is Logic VI* (Blaha Research, Auburn, NH, 2016).
______, 2016c, *New Boson Quantum Field Theory, Dark Matter Dynamics, Dark Matter Fermion Layer Mixing, Genesis of Higgs Particles, New Layer Higgs Masses, Higgs Coupling Constants, Non-Abelian Higgs Gauge Fields, Physics is Logic VII* (Blaha Research, Auburn, NH, 2016).
______, 2016d, *Unification of the Strong Interactions and Gravitation: Quark Confinement Linked to Modified Short-Distance Gravity; Physics is Logic VIII* (Blaha Research, Auburn, NH, 2016).
______, 2016e, *MoND: Unification of the Strong Interactions and Gravitation II, Quark Confinement Linked to Large-Scale Gravity, Physics is Logic IX* (Blaha Research, Auburn, NH, 2016).
______, 2016f, *CQ Mechanics: A Unification of Quantum & Classical Mechanics, Quantum/Semi-Classical Entanglement, Quantum/Classical Path Integrals, Quantum/Classical Chaos* (Blaha Research, Auburn, NH, 2016).
______, 2016g, *GEMS Unified Gravity, ElectroMagnetic and Strong Interactions: Manifest Quark Confinement, A Solution for the Proton Spin Puzzle, Modified Gravity on the Galactic Scale* (Pingree Hill Publishing, Auburn, NH, 2016).
______, 2016h, *Unification of the Seven Boson Interactions based on the Riemann-Christoffel Curvature Tensor* (Pingree Hill Publishing, Auburn, NH, 2016).
______, 2017a, *Unification of the Eleven Boson Interactions based on 'Rotations of Interactions'* (Pingree Hill Publishing, Auburn, NH, 2017).
______, 2017b, *The Origin of Fermions and Bosons, and Their Unification* (Pingree Hill Publishing, Auburn, NH, 2017).
______, 2017c, *Megaverse: The Universe of Universes* (Pingree Hill Publishing, Auburn, NH, 2017).
______, 2017d, *SuperSymmetry and the Unified SuperStandard Model* (Pingree Hill Publishing, Auburn, NH, 2017).

______, 2017e, *From Qubits to the Unified SuperStandard Model with Embedded SuperStrings: A Derivation* (Pingree Hill Publishing, Auburn, NH, 2017).
______, 2017f, *The Unified SuperStandard Model in Our Universe and the Megaverse: Quarks, ...* , (Pingree Hill Publishing, Auburn, NH, 2017).
______, 2018a, *The Unified SuperStandard Model and the Megaverse SECOND EDITION A Deeper Theory based on a New Particle Functional Space that Explicates Quantum Entanglement Spookiness (Volume 1)* (Pingree Hill Publishing, Auburn, NH, 2018).
______, 2018b, *Cosmos Creation: The Unified SuperStandard Model, Volume 2, SECOND EDITION* (Pingree Hill Publishing, Auburn, NH, 2018).
______, 2018c, *God Theory (*Pingree Hill Publishing, Auburn, NH, 2018).
______, 2018d, *Immortal Eye: God Theory: Second Edition* (Pingree Hill Publishing, Auburn, NH, 2018).
______, 2018e, *Unification of God Theory and Unified SuperStandard Model THIRD EDITION* (Pingree Hill Publishing, Auburn, NH, 2018).
______, 2019a, *Calculation of: QED α = 1/137, and Other Coupling Constants of the Unified SuperStandard Theory* (Pingree Hill Publishing, Auburn, NH, 2019).
______, 2019b, *Coupling Constants of the Unified SuperStandard Theory SECOND EDITION* (Pingree Hill Publishing, Auburn, NH, 2019).
______, 2019c, *New Hybrid Quantum Big_Bang–Megaverse_Driven Universe with a Finite Big Bang and an Increasing Hubble Constant* (Pingree Hill Publishing, Auburn, NH, 2019).
______, 2019d, *The Universe, The Electron and The Vacuum* (Pingree Hill Publishing, Auburn, NH, 2019).
______, 2019e, *Quantum Big Bang – Quantum Vacuum Universes (Particles)* (Pingree Hill Publishing, Auburn, NH, 2019).
______, 2019f, *The Exact QED Calculation of the Fine Structure Constant Implies ALL 4D Universes have the Same Physics/Life Prospects* (Pingree Hill Publishing, Auburn, NH, 2019).
______, 2019g, *Unified SuperStandard Theory and the SuperUniverse Model: The Foundation of Science* (Pingree Hill Publishing, Auburn, NH, 2019).
______, 2020a, *Quaternion Unified SuperStandard Theory (The QUeST) and Megaverse Octonion SuperStandard Theory (MOST)* (Pingree Hill Publishing, Auburn, NH, 2020).
______, 2020b, *United Universes Quaternion Universe - Octonion Megaverse* (Pingree Hill Publishing, Auburn, NH, 2020).
______, 2020c, *Unified SuperStandard Theories for Quaternion Universes & The Octonion Megaverse* (Pingree Hill Publishing, Auburn, NH, 2020).
______, 2020d, *The Essence of Eternity: Quaternion & Octonion SuperStandard Theories* (Pingree Hill Publishing, Auburn, NH, 2020).
______, 2020e, *The Essence of Eternity II* (Pingree Hill Publishing, Auburn, NH, 2020).
______, 2020f, *A Very Conscious Universe* (Pingree Hill Publishing, Auburn, NH, 2020).
______, 2020g, *Hypercomplex Universe* (Pingree Hill Publishing, Auburn, NH, 2020).
______, 2020h, *Beneath the Quaternion Universe* (Pingree Hill Publishing, Auburn, NH, 2020).
______, 2020i, *Why is the Universe Real? From Quaternion & Octonion to Real Coordinates* (Pingree Hill Publishing, Auburn, NH, 2020).
______, 2020j, *The Origin of Universes: of Quaternion Unified SuperStandard Theory (QUeST); and of the Octonion Megaverse (UTMOST)* (Pingree Hill Publishing, Auburn, NH, 2020).
______, 2020k, *The Seven Spaces of Creation: Octonion Cosmology* (Pingree Hill Publishing, Auburn, NH, 2020).
______, 2020l, *From Octonion Cosmology to the Unified SuperStandard Theory of Particles* (Pingree Hill Publishing, Auburn, NH, 2020).
______, 2021a, *Pioneering the Cosmos* (Pingree Hill Publishing, Auburn, NH, 2021).
______, 2021b, *Pioneering the Cosmos II* (Pingree Hill Publishing, Auburn, NH, 2021).
______, 2021c, *Beyond Octonion Cosmology* (Pingree Hill Publishing, Auburn, NH, 2021).
______, 2021d, *Universes are Particles* (Pingree Hill Publishing, Auburn, NH, 2021).
______, 2021e, *Octonion-like dna-based life, Universe expansion is decay, Emerging New Physics* (Pingree Hill Publishing, Auburn, NH, 2021).
______, 2021f, *The Science of Creation New Quantum Field Theory of Spaces* (Pingree Hill Publishing, Auburn, NH, 2021).
______, 2021g, *Quantum Space Theory With Application to Octonion Cosmology & Possibly To Fermionic Condensed Matter* (Pingree Hill Publishing, Auburn, NH, 2021).
______, 2021h, *21st Century Natural Philosophy of Octonion Cosmology , and Predestination, Fate, and Free Will* (Pingree Hill Publishing, Auburn, NH, 2021).

______, 2021i, *Beyond Octonion Cosmology II : Origin of the Quantum; A New Generalized Field Theory (GiFT); A Proof of the Spectrum of Universes; Atoms in Higher Universes* (Pingree Hill Publishing, Auburn, NH, 2021).
______, 2021j, *Integration of General Relativity and Quantum Theory: Octonion Cosmology, GiFT, Creation/Annihilation Spaces CASe, Reduction of Spaces to a Few Fermions and Symmetries in Fundamental Frames* (Pingree Hill Publishing, Auburn, NH, 2021).
______, 2022a, *New View of Octonion Cosmology Based on the Unification of General Relativity and Quantum Theory* (Pingree Hill Publishing, Auburn, NH, 2022).
______, 2022b, *The Dust Beneath Hypercomplex Cosmology* (Pingree Hill Publishing, Auburn, NH, 2022).
______, 2022c, *Passing Through Nature to Eternity: ProtoCosmos, HyperCosmos, Unified SuperStandard Theory* (Pingree Hill Publishing, Auburn, NH, 2022).
______, 2022d, *HyperCosmos Fractionation and Fundamental Reference Frame Based Unification: Particle Inner Space Basis of Parton and Dual Resonance Models* (Pingree Hill Publishing, Auburn, NH, 2022).
______, 2022e, *A New UniDimension ProtoCosmos and SuperString F-Theory Relation to the HyperCosmos* (Pingree Hill Publishing, Auburn, NH, 2022).
______, 2022f, *The Cosmic Panorama: ProtoCosmos, HyperCosmos, Unified SuperStandard Theory (UST) Derivation* (Pingree Hill Publishing, Auburn, NH, 2022).
_____, 2022g, *Ultimate Origin: ProtoCosmos and HyperCosmos* (Pingree Hill Publishing, Auburn, NH, 2022).
______, 2023a, *UltraUnification and the Generation of the Cosmos* (Pingree Hill Publishing, Auburn, NH, 2023).
______, 2023b, *God and and Cosmos Theory* (Pingree Hill Publishing, Auburn, NH, 2023).
______, 2023c, *A New Completely Geometric SU(8) Cosmos Theory; New PseudoFermion Fields; Fibonacci-like Dimension Arrays; Ramsey Number Approximation* (Pingree Hill Publishing, Auburn, NH, 2023).
______, 2023d, *Newton's Apple is Now the Fermion* (Pingree Hill Publishing, Auburn, NH, 2023).
______, 2023e,*Cosmos Theory: The Sub-Particle Gambol Model* (Pingree Hill Publishing, Auburn, NH, 2023).
_____, 2024a, *Cosmos-Universe-Particle-Gambol Theory* (Pingree Hill Publishing, Auburn, NH, 2024).
_____, 2024b, *Fractal Cosmos Theory* (Pingree Hill Publishing, Auburn, NH, 2024).
_____, 2024c, *Fractal Cosmic Curve: Tensor-Based CosmosTheory* (Pingree Hill Publishing, Auburn, NH, 2024).
_____, 2024d, *The Eternal Form of Cosmos Theory* (Pingree Hill Publishing, Auburn, NH, 2024).
_____, 2024e, *The Eternal Form of Cosmos Theory Third Edition* (Pingree Hill Publishing, Auburn, NH, 2024).
_____, 2024f, *Fundamental Constants of Cosmos Theory and The Standard Model* (Pingree Hill Publishing, Auburn, NH, 2024).
_____, 2024g, *Quark, Lepton, W and Z Masses of Cosmos Theory and The Standard Model* (Pingree Hill Publishing, Auburn, NH, 2024).
_____, 2024h, *Geometric Cosmos Geometric Universe* (Pingree Hill Publishing, Auburn, NH, 2024).
_____, 2024i, *Particles and Universes of Cosmos Theory* (Pingree Hill Publishing, Auburn, NH, 2024).
_____, 2024j, *Unification of the Subluminal and the Superluminal in Cosmos Theory* (Pingree Hill Publishing, Auburn, NH, 2024).
_____, 2024k, *The Dawn of Dynamic Cosmos Dimension Arrays* (Pingree Hill Publishing, Auburn, NH, 2024).
_____, 2024ℓ, *Structure and Dynamics of Cosmos Theory and the Unified SuperStandard Theory* (Pingree Hill Publishing, Auburn, NH, 2024).
_____, 2025a, *Black Holes, White Holes, and Superluminal Starship* (Pingree Hill Publishing, Auburn, NH, 2025).
_____, 2025b, *The Sequence Theory Foundation of Cosmos Theory and Gambol Theory* (Pingree Hill Publishing, Auburn, NH, 2025).
____, 2025c, *BEYOND TIME AND SPACE Fermion Strata & Consistency Conditions for Sequences, Cosmos Spaces and Gambols* (Pingree Hill Publishing, Auburn, NH, 2025).
____, 2025d, *The C o s m o s as a Physics Organism* (Blaha Research, Auburn, NH, 2025).
____, 2025e, *The Root of Reality* (Blaha Research, Auburn, NH, 2025).
____, 2025f, *The Fractal Cosmos* (Blaha Research, Auburn, NH, 2025).
____, 2025g, *The Fractal Cosmos II* (Blaha Research, Auburn, NH, 2025).
____, 2025h, *The Fractal Cosmos III* (Blaha Research, Auburn, NH, 2025).
____, 2025i, *Quantum Coalescent Cosmos Fractal Cosmos IV* (Blaha Research, Auburn, NH, 2025).
____, 2025j, *The Entropic Basis of Cosmos Theory The Agent of Origin* (Blaha Research, Auburn, NH, 2025).
Eddington, A. S., 1952, *The Mathematical Theory of Relativity* (Cambridge University Press, Cambridge, U.K., 1952).
Fant, Karl M., 2005, *Logically Determined Design: Clockless System Design With NULL Convention Logic* (John Wiley and Sons, Hoboken, NJ, 2005).
Feinberg, G. and Shapiro, R., 1980, *Life Beyond Earth: The Intelligent Earthlings Guide to Life in the Universe* (William Morrow and Company, New York, 1980).

Gelfand, I. M., Fomin, S. V., Silverman, R. A. (tr), 2000, *Calculus of Variations* (Dover Publications, Mineola, NY, 2000).
Giaquinta, M., Modica, G., Souchek, J., 1998, *Cartesian Coordinates in the Calculus of Variations* Volumes I and II (Springer-Verlag, New York, 1998).
Giaquinta, M., Hildebrandt, S., 1996, *Calculus of Variations* Volumes I and II (Springer-Verlag, New York, 1996).
Gradshteyn, I. S. and Ryzhik, I. M., 1965, *Table of Integrals, Series, and Products* (Academic Press, New York, 1965).
Heitler, W., 1954, *The Quantum Theory of Radiation* (Claendon Press, Oxford, UK, 1954).
Huang, Kerson, 1992, *Quarks, Leptons & Gauge Fields 2nd Edition* (World Scientific Publishing Company, Singapore, 1992).
Jost, J., Li-Jost, X., 1998, *Calculus of Variations* (Cambridge University Press, New York, 1998).
Kaku, Michio, 1993, *Quantum Field Theory*, (Oxford University Press, New York, 1993).
Kirk, G. S. and Raven, J. E., 1962, *The Presocratic Philosophers* (Cambridge University Press, New York, 1962).
Landau, L. D. and Lifshitz, E. M., 1987, *Fluid Mechanics 2nd Edition*, (Pergamon Press, Elmsford, NY, 1987).
Rescher, N., 1967, *The Philosophy of Leibniz* (Prentice-Hall, Englewood Cliffs, NJ, 1967).
Riesz, Frigyes and Sz.-Nagy, Béla, 1990, *Functional Analysis* (Dover Publications, New York, 1990).
Sakurai, J. J., 1964, *Invariance Principles and Elementary Particles* (Princeton University Press, Princeton, NJ, 1964).
Weinberg, S., 1972, *Gravitation and Cosmology* (John Wiley and Sons, New York, 1972).
Weinberg, S., 1995, *The Quantum Theory of Fields Volume I* (Cambridge University Press, New York, 1995).

INDEX

About the Author

Stephen Blaha is a well-known Physicist and Man of Letters with interests in Science, Society and civilization, the Arts, and Technology. He had an Alfred P. Sloan Foundation scholarship in college. He received his Ph.D. in Physics from Rockefeller University. He has served on the faculties of several major universities. He was also a Member of the Technical Staff at Bell Laboratories, a manager at the Boston Globe Newspaper, a Director at Wang Laboratories, and President of Blaha Software Inc. and of Janus Associates Inc. (NH).

Among other achievements he was a co-discoverer of the "r potential" for heavy quark binding developing the first (and still the only demonstrable) non-Aeolian gauge theory with an "r" potential; first suggested the existence of topological structures in superfluid He-3; first proposed Yang-Mills theories would appear in condensed matter phenomena with non-scalar order parameters; first developed a grammar-based formalism for quantum computers and applied it to elementary particle theories; first developed a new form of quantum field theory without divergences (thus solving a major 60 year old problem that enabled a unified theory of the Standard Model and Quantum Gravity without divergences to be developed); first developed a formulation of complex General Relativity based on analytic continuation from real space-time; first developed a generalized non-homogeneous Robertson-Walker metric that enabled a quantum theory of the Big Bang to be developed without singularities at t = 0; first generalized Cauchy's theorem and Gauss' theorem to complex, curved multi-dimensional spaces; received Honorable Mention in the Gravity Research Foundation Essay Competition in 1978; first developed a physically acceptable theory of faster-than-light particles; first derived a composition of extremums method in the Calculus of Variations; first quantitatively suggested that inflationary periods in the history of the universe were not needed; first proved Gödel's Theorem implies Nature must be quantum; provided a new alternative to the Higgs Mechanism, and Higgs particles, to generate masses; first showed how to resolve logical paradoxes including Gödel's Undecidability Theorem by developing Operator Logic and Quantum Operator Logic; first developed a quantitative harmonic oscillator-like model of the life cycle, and interactions, of civilizations; first showed how equations describing superorganisms also apply to civilizations. A recent book shows his theory applies successfully to the past 14 years of history and to *new* archaeological data on Andean and Mayan civilizations as well as Early Anatolian and Egyptian civilizations.

He first developed an axiomatic derivation of the form of The Standard Model from geometry – space-time properties – The Unified SuperStandard Model. It unifies all the known forces of Nature. It also has a Dark Matter sector that includes a Dark ElectroWeak sector with Dark doublets and Dark gauge interactions. It uses quantum coordinates to remove infinities that crop up in most interacting quantum field theories and additionally to remove the infinities that appear in the Big Bang and generate inflationary growth of the universe. It shows gravity has a MOND-like form without sacrificing Newton's Laws. It relates the interactions of the MOND-like sector of gravity with the r-potential of Quark Confinement. The axioms of the theory lead to the question of their origin. We suggest in the preceding edition of this book it can be attributed to an entity with God-like properties. We explore these properties in "God Theory" and show they predict that the Cosmos exists forever although individual universes (or incarnations of our universe) "come and go." Several other important results emerge from God Theory such a functionally triune God. The Unified SuperStandard Theory has many other important parts described in the Current Edition of *The Unified SuperStandard Theory* and expanded in subsequent volumes.

Blaha has had a major impact on a succession of elementary particle theories: his Ph.D. thesis (1970), and papers, showed that quantum field theory calculations to all orders in ladder approximations could not give scaling deep inelastic electron-nucleon scattering. He later showed the eigenvalue equation for the fine structure constant α in Johnson-Baker-Willey QED had a zero at $\alpha = 1$ not 1/137 by solving the Schwinger-Dyson equations to all orders in an approximation that agreed with exact results to 4th order in α thus ending interest in this theory. In 1979 at Professor Ken Johnson's (MIT) suggestion he calculated the proton-neutron mass difference in the MIT bag model and found the result had the wrong sign reducing interest in the bag model. These results all appear in Physical Review papers. In the 2000's he repeatedly pointed out the shortcomings of SuperString theory and showed that The Standard Model's form could be derived from space-time geometry by an extension of Lorentz transformations to faster than light transformations. This deeper space-time basis greatly increases the possibility that it is part of THE fundamental theory. Recently, Blaha showed that the Weak interactions differed significantly from the Strong, electromagnetic and gravitation interactions in important respects while these interactions had similar features, and suggested that ElectroWeak theory, which is essentially a glued union of the Weak interactions and Electromagnetism, possibly modulo unknown Higgs particle features, be replaced by a unified theory of the other interactions combined with a stand-alone Weak interaction theory. Blaha also showed that, if Charmonium calculations are taken seriously, the Strong interaction coupling constant is only a factor of five larger than the electromagnetic coupling constant, and thus Strong interaction perturbation theory would make sense and yield physically meaningful results.

In graduate school (1965-71) he wrote substantial papers in elementary particles and group theory: The Inelastic E- P Structure Functions in a Gluon Model. Phys. Lett. B40:501-502,1972; Deep-Inelastic E-P Structure Functions In A Ladder Model With Spin 1/2 Nucleons, Phys.Rev. D3:510-523,1971; Continuum Contributions To The Pion Radius, Phys. Rev. 178:2167-2169,1969; Character Analysis of U(N) and SU(N), J. Math. Phys. 10, 2156 (1969); and The Calculation of the Irreducible Characters of the Symmetric Group in Terms of the Compound Characters, (Published as Blaha's Lemma in D. E. Knuth's book: *The Art of Computer Programming Vols. 1 – 4*).

In the early 1980's Blaha was also a pioneer in the development of UNIX for financial, scientific and Internet applications: benchmarked UNIX versions showing that block size was critical for UNIX performance, developing financial modeling software, starting database benchmarking comparison studies, developing Internet-like UNIX networking (1982) and developing a hybrid shell programming technique (1982) that was a precursor to the PERL programming language. He was also the manager of the AT&T ten-year future products development database. His work helped lead to commercial UNIX on computers such as Sun Micros, IBM AIX minis, and Apple computers.

In the 1980's he pioneered the development of PC Desktop Publishing on laser printers and was nominated for three "Awards for Technical Excellence" in 1987 by PC Magazine for PC software products that he designed and developed.

Recently he has developed a theory of Megaverses – actual universes of which our universe is one – with quantum particle-like properties based on the Wheeler-DeWitt equation of Quantum Gravity. He has developed a theory of a baryonic force, which had been conjectured many years ago, and estimated the strength of the force based on discrepancies in measurements of the gravitational constant G. This force, operative in D-dimensional space, can be used to escape from our universe in "uniships" which are the equivalent of the faster-than-light starships proposed in the author's earlier books. Thus travel to other universes, as well as to other stars is possible.

Blaha also considered the complexified Wheeler-DeWitt equation and showed that its limitation to real-valued coordinates and metrics generated a Cosmological Constant in the Einstein equations.

The author has also recently written a series of books on the serious problems of the United States and their solution as well as a book on the decline of Mankind that will follow from current social and genetic trends in Mankind.

In the past twenty years Dr. Blaha has written over 80 books on a wide range of topics. Some recent major works are: *From Asynchronous Logic to The Standard Model to Superflight to the Stars*, *All the Universe!*, *SuperCivilizations: Civilizations as Superorganisms, America's Future: an Islamic Surge, ISIS, al Qaeda, World Epidemics, Ukraine, Russia-China Pact, US Leadership Crisis, The Rises and Falls of Man – Destiny – 3000 AD: New Support for a Superorganism MACRO-THEORY of CIVILIZATIONS From CURRENT WORLD TRENDS and NEW Peruvian, Pre-Mayan, Mayan, Anatolian, and Early Egyptian Data, with a Projection to 3000 AD,* and *Mankind in Decline: Genetic Disasters, Human-Animal Hybrids, Overpopulation, Pollution, Global Warming, Food and Water Shortages, Desertification, Poverty, Rising Violence, Genocide, Epidemics, Wars, Leadership Failure.*

He has taught approximately 4,000 students in undergraduate, graduate, and postgraduate corporate education courses primarily in major universities, and large companies and government agencies.

He developed a quantum theory, The Unified SuperStandard Theory (UST), which describes elementary particles in detail without the difficulties of conventional quantum field theory. He found that the internal symmetries of this theory could be exactly derived from an octonion theory called QUeST. He further found that another octonion theory (UTMOST) describes the Megaverse. It can hold QUeST universes such as our own universe. It has an internal symmetry structure which is a superset of the QUeST internal symmetries.

Recently he developed Octonion Cosmology. He replaced it with HyperCosmos theory, which has significantly better features. He developed a fractionalization process for dimensions, particles and symmetry groups. He also described transformation that reduced particles and dimensions to a far more compact form. He also developed a precursor theory ProtoCosmos that leads to the HyperCosmos.

The author showed that space-time and Internal Symmetries can be unified in any of the ten HyperCosmos spaces in their associated HyperUnification spaces. The combined set of HyperUnification spaces enable all HyperCosmos dimensions to be obtained by a General Relativistic transformation from one primordial dimension in the 42 space-time dimension unified HyperUnification space.

At present the author devel;oped the Cosmos Theory that incorporates ProtoCosmos Theory, HyperCosmos Theory, Limos Theory, Second Kind HyperCosmos Theory and HyperUnification Spaces. He has introduced PseudoFermion wave functions and theory, He has related Cosmos Theory to Regge trajectories of spaces, parton theory, Veneziano amplitudes, Fibonacci numbers and Ramsey numbers. He has calculated an approximation to the difficult R(n,n) Ramsey numbers.

He has developed a Gambol Model that successfully accounts for e-p deep inelastic scattering, fundamental particle resonances, hadron scattering, and the inner structure of particles based on confinement through Casimir forces of ideal gambol gases. The Gambol Planckian Distribution was derived.

He has applied the Gambol Model to particles, universes, and the Cosmos of universes. He showed that the Cosmos may have a distribution of 23 universes corresponding to various Cosmos spaces.

Recently he showed that Cosmos Theory follows from the number of independent asymmetric tensors in a dimension r. He also showed the close parallel between the form of γ-matrices and Cosmos Theory dimension arrays. The closeness suggested that dimension arrays have the same importance as γ-matrices for fermions.

He demonstrated that the pressure of fermions within a space of dimension r balances the Casimir vacuum energy force for 18 dimensions. He showed that $2e\pi = 17.02$ marks the critical point where pressure balances Casimir force, which implies $r = 18$ is the highest dimension Physical Cosmos space. The dimension $2e\pi$ appears to set the approximate dimension for Cosmos spaces with dimension array size $2^{r+4} \cong (17.02/8)^{r+4}. \cong (e\pi/4)^{r+4}. \cong 2.13^{r+4}$.

Now he has found the sequences of Coupling Constant values and fermion masses in the Standard Model and UST. He unified the Cosmos spaces spectrum, the coupling constant spectrum and the fermion mass sequences in a fundamental scaling differential equation.

Cosmos Theory is now well explored.